CONNECTING NEIGHBORS IN AMERICA'S OUTBACK

by Terry A. Del Bene

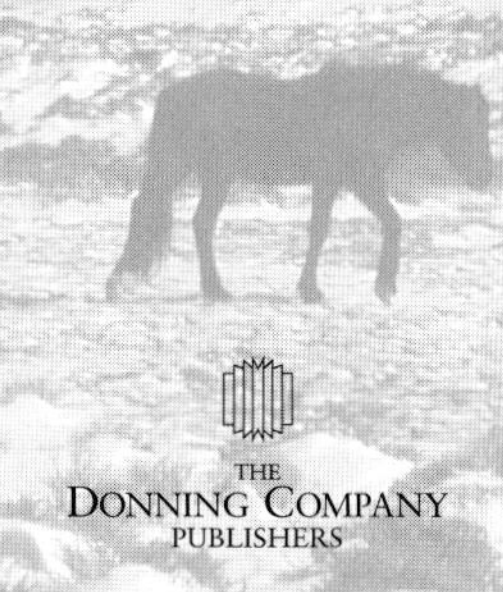

The Donning Company Publishers
184 Business Park Drive, Suite 206
Virginia Beach, VA 23462

THE
DONNING COMPANY
PUBLISHERS

Lex Cavanah, General Manager
Nathan Stufflebean, Production Supervisor
Philip Briscoe, Editor
Jeremy Glanville, Graphic Designer
Monika Ebertz, Imaging Artist

Cathleen Norman, Project Director

Library of Congress Cataloging-in-Publication Data

Del Bene, Terry Alan, author.
A phone where the buffalo roamed : connecting neighbors in America's outback / by Terry A. Del Bene.
pages cm
Includes bibliographical references.
ISBN 978-1-68184-001-7 (hardcover : alk. paper)
1. Union Telephone Company. 2. Telephone companies--Wyoming--History. I. Title.
HE8846.U539D45 2015
384.606'5787--dc23

2015033362

Printed in the United States of America at Walsworth

TABLE OF CONTENTS

A Phone on the Range 5

Acknowledgments 6

Preface 8

Introduction 20

CHAPTER 1
Lone Star State Tribulations and Cowboy State Opportunities 24

CHAPTER 2
The First Generation:
John D. Woody's Dream 46

CHAPTER 3
The Second Generation:
Making the Dream Come True 84

CHAPTER 4
The Third Generation:
Leaps of Faith 148

CHAPTER 5
The Fourth Generation:
Technological Horizons Within a Changing Industrial Landscape 186

CHAPTER 6
The Fifth Generation:
Keeping the Dream Alive 208

APPENDICES:
Union Telephone Board of Directors 216
Union Telephone Highlights: A Hundred-and-One Years in a Glimpse 219

REFERENCES 224

Neither Union Telephone nor I would be where we are today were it not for the vision of my father. This book is dedicated to my dad, John D. Woody.

—Howard Woody

A PHONE ON THE RANGE

(Sung to the tune of the 1876 song "Home on the Range"
by Dr. Brewster M. Higley)

Oh, give me a phone,

Where the buffalo roam,

Where the landlines rest under the hay,

Where microwaves fly

O'er snowy mountains nearby

And carry me all that you say.

A solitary wild horse walks through a landscape typical of those serviced by Union Telephone.
Photograph courtesy of Terry Del Bene

ACKNOWLEDGMENTS

I acknowledge the help of the many individuals and organizations that have made it possible to prepare this book within a short time frame. I am tremendously grateful for all of the assistance they have provided.

Linda Montoya has made substantial contributions to this book by arranging and taping interviews, providing documents, answering questions, and being the go-to person at Union Telephone Company. Thank you, Linda, for all the great help.

The company and family-member history comes, in large part, from scores of interviews recorded by Linda Montoya, Christine Hinckley, Ruth Barclay, and me. Thank you to everyone who agreed to provide interviews. This includes Bonnielee Woody Alexander, Stacey Aughe, Ruth Barclay, Joan Jane Byrne, Tom Davidow, Carol Davis, Louise Davis, Geoffrey Dutray, Jan Fasselin, Alan Jaeger, Joyce Garris, Rae Geibel, Kindra Green, Kim Halford, Cynthia Adams Harrison, Leslie F. Henderson, Ken King, Roger Lind, Betty Jo Lyman, Linda Montoya, Katie Olsen, Dick Perue, Walter Riebenack, Bonnie J. Shannon, Raymond Tanner, Betty M. Vyvey, Vance Walker, Abbey Woody, Eric Jon Woody, Gene Woody, Howard D. Woody, James H. Woody, John G. Woody, Michael Woody, Pam Woody, Robert Brian Woody, and Ryan Woody. With so much good material, it was difficult choosing what to include in this book.

Tom Davidow and Cindy Adams Harrison have worked as hard as anyone to make this book a reality. Thank you to both of you for the long hours of interviews and conference calls.

Source materials and the access to collections were provided by a number of organizations, including Union Telephone Company, Sweet Home Genealogical Society, Grand Encampment Museum, Sweetwater County Museum, Uinta County Museum, Saratoga Museum, Wyoming State Archives, and the Public Utilities Commission of Wyoming. Candy Moulton was her usual helpful self in providing access to source materials.

The Union Telephone staff, including Jan Fasselin and many others, helped provide images. Based upon verbal descriptions, Mike Jackson created the illustration of the snow plane.

It might surprise the reader to know that detailed board-meeting minutes for the Union Telephone Company do not exist for the first forty-two years of the corporation's existence. Accordingly, the history of those years relies heavily upon the imperfect memories of those who "lived to tell the tale." As a result, there are some events described herein where a range of dates is the best that can be determined. In themselves, board-meeting minutes are another form of imperfect recollection and can raise as many issues as they resolve. Thank you to John G. Woody, Linda Montoya, and Ruth Barclay for delving into the minutes when I asked questions about the timing of events.

Thank you to my wife, Penelope, for tolerating my absences while gathering materials for the preparation of the manuscript, as well as my all-too-frequent "lock-downs" in my office during its creation.

Most of all I would like to thank the Woody family for trusting me to weave the loose strands of multiple story lines into a tapestry of their family history. Those who have never been engaged in trying to pull together an accurate history might not realize that recounting tragedies and difficult moments in our lives is not an easy thing to do. Many people have unpleasant memories they tuck away. Being forced to revisit those memories can be difficult. However, and to their credit, the Woody family commissioned this book as a history of the people involved, and not merely a recounting of the company's journey on its path to success. Those hard moments reveal far more about the character of the family than ledger-sheet statics can provide. It takes an act of courage to trust an outsider to do justice to a family's view of its past.

Terry A. Del Bene

PREFACE

Things may come to those who wait, but only the things left by those who hustle.

—Abraham Lincoln

It has been one of Wyoming's best-kept secrets that the centennial year of the Union Telephone Company was 2014. Accordingly, the telephone company (which is headquartered in Mountain View, Uinta County, Wyoming) is one of the oldest continuously operated, family-owned businesses in Wyoming. Between eighty and ninety percent of all businesses in North America are family owned. Union Telephone is among the few of these companies that survived to celebrate their fiftieth anniversary, much less double that time. In recognition of their remarkable achievement in traveling time, the Union Telephone Company has commissioned this remembrance of those who made the journey possible.

Union Telephone's odyssey began in 1914 when John D. Woody incorporated the company under the laws of the state of Wyoming. The company's founder intended to create a utility that linked neighbors together and provided them with service of as high of a quality as was available in more populated areas. For John D. Woody, this was not just a business model—it was a vision and a calling. This dream became John D. Woody's purpose in life, and a mission he bequeathed to his descendants.

Union Telephone Company has grown from a shoestring assemblage of ranch lines to a multimillion-dollar corporation, serving subscribers in several Rocky Mountain states and linking customers throughout the United States and in several foreign countries. The company started out as a homespun entity that used homemade technology, and it has evolved into a state-of-the art service provider that is at the forefront of establishing standards for the industry. It has been a remarkable journey.

Telephone companies in several other states, such as Wisconsin, New Hampshire, Ohio, and New Jersey, share the Union Telephone Company name. The Union Telephone Company of Wyoming was not the first company to use that name, as there was a Union Telephone Company in the Midwest as early as 1902. To make the matter more confusing, there are two communities named Mountain View in Wyoming. Happily, the Mountain View in Natrona County did not become home for one of those other Union Telephone companies, and there is but one company by that name in Wyoming.

The Union Telephone story is a tale of generations. Each generation of the Woody family has brought different talents and personalities to the business, and all of them have remained true to the common dream. Five generations of the Woody family have worked to maintain the vision of the company's founder. A sixth generation is waiting for the day when they will join the business and keep both the Union Telephone and Woody family traditions alive.

This book is divided into chapters that focus on the contributions of each generation. To tell this tale as a multigenerational story, the narrative is forced to jump back and forth in time during certain portions of the story. There were no clean breaks between the accomplishments of the respective generations. Excluding the first years, Union Telephone has been a multigenerational family business. A timeline appears in the appendices in order to provide the reader with a chronology of the significant events described herein.

In Union Telephone's centennial year, second-generation member Howard D. Woody celebrated his ninety-second birthday. Howard, the only son of the company's founder (and whose wife served the company as an operator), was fittingly born in a tiny room behind the Union Telephone switchboard. Clearly destined to be part of the company, Howard D. Woody has worked at Union since he was a child, helping his father inspect lines at the tender age of eight. After eighty-four years of involvement in the business, Howard continues to guide the company despite his "retirement."

In many respects, Howard has become the soul of the company and the loving keeper of his father's dream. Though the day-to-day operations of the company are being run by later generations, Howard shows up to work almost every day and still serves as a grand mentor for the younger members of the family. So much of the story of the success of Union Telephone is Howard's, but to tell only his story is to miss the true narrative of this remarkable company and family. At the time of Union Telephone's centennial, the only generations uninvolved with the company were generations one and six. Generations two through five are making things happen every day at Union Telephone, which was rechristened as Union Cellular almost twenty-five years ago, and Union Wireless nine years before the centennial year.

As with all family businesses, the issue of how to weave multiple generations into a corporation's structure and leadership represents one of the most vexing issues a kin group has to face. Deep and lasting rifts between family members can result from the day-to-day business decisions, sometimes over the most trivial of matters. Families who recognize that decisions in a family business are both family and corporate decisions tend to do better at resolving internal conflicts. Sometimes it takes a few stumbles to figure out how best to balance family and corporate issues. Often, finding this balance requires outside help. Ultimately, the families that stick together during times of crisis or challenge seem to best weather the inevitable storms. When that horse throws you, it's best to dust yourself off and get back in the saddle. Bruises can be learning tools, and they remind all of us to pay attention and hang on.

The Union Telephone story is a classic tale of Wyoming, a state known for having dreams of success swallowed up in its vast expanses, blown away by its constant winds, buried in its icy snow drifts, mired in its muddy roads, charred in its wild fires, shattered by petulant neighbors, and/or withered for lack of water or investment dollars. Wyoming does not possess a reputation for being a place to go in order to have one's dreams come true. The oftentimes harsh environment of the Cowboy State is a place to test one's resolve and ability to take on challenges. Dreams do come true in such a place, but realizing them takes effort, commitment, and sometimes plain old good luck. As my grandfather once reminded me, "Our dreams are irrigated with nothing more than our own sweat." That could very easily be a motto for the owners of the Union Telephone Company, the Woody family.

What follows is not your typical "rags-to-riches" fable. For roughly fifty years, the Woody family's reward for working hard to create their telephone company in the West's Rocky Mountains was red ink and seemingly endless backbreaking work within one of the harshest climates on the continent. As it was earned, the revenue from the company was plowed back into the business, leaving the shareholders with virtually no dividends or earnings. Throughout three and a half lean decades, the Woody family refused to admit defeat, and only offered the company for sale due to the advanced years of its founder. Even after that brush with liquidation, the family held on for almost the next fifteen years without profits. After reaching profitability, the company's struggle to stay in the black continued through decade after decade, as the family increased both in number and in their successes in one of the most competitive industries.

Building something as fragile and as large as a communications system in the cruel environment of the Rocky Mountains was no easy task. The Woody family has evidenced something akin to the patience of the land itself, a survival technique that works well in Wyoming. Some make lemonade when life hands them lemons, and some plant lemon trees and wait for an entire orchard to grow. The Woody family has mastered planting lemon trees, and they have made the sparse, high-desert basins of Wyoming bloom, bringing modern telecommunications to landscapes that in many ways remain every bit as isolated and foreboding as they were in the nineteenth century.

Wyomingites recognize that the commitments between neighbors are not written in corporate ledger books. They are written somewhere in the heart. What is a dream worth? The Woody family appears to have answered, "More than money, and much more than comfort." Generation upon generation of this remarkable family took up the seemingly insane dream of the company's founder and made it a reality. Each generation has taken a patently hands-on approach to running the company and immersing themselves in learning whatever is necessary to make telecommunications systems operate in America's Outback. In the Woody family, there are no desk jockeys that allow others to do all the hard work. When there is work to be done, the family members roll up their collective sleeves and pitch in.

The Woody family did not idly sit by and wait for good fortune to arrive at its own pace. Family members built the company pole by pole, subscriber by subscriber, and exchange by exchange throughout decades of incremental growth. This deed could not have been done without help from neighbors, as well as timely assistance from a government that aided small telephone companies in rural America. Friends are important, but good neighbors are essential in the wind-swept mountains and basins of Wyoming. And a little good luck to soften the all-too-frequent runs of bad luck didn't hurt, either. Lady Luck has been good to the company, but only because the Woody family established the conditions that gave her a chance to show what she could do.

Unlike Robert MacDougall's *The People's Network: The Political Economy of the Telephone in the Gilded Age*, this centennial history of Union Telephone Company is not intended to be a scholarly treatise on the development of rural telephone systems. For the sake of readability, the use of references and notes will be kept to a minimum. The general history inserted in the footnotes is well known, but it acts as a way of providing context and is important to understanding how the circumstances in Wyoming and the world impacted both the company and the family.

In writing a book about multiple generations of the same family, it is not unusual for individuals in different generations to have similar names. In this book, there are two John Woodys who are major actors in this story. John D. Woody founded the company. John G. Woody served as general manager and held a variety of jobs in the company during much of its history, and in the centennial year remained critical to the operation of the company. The former is discussed herein as John D., and the latter as John G. This convention is employed with apologies to both Johns, neither of whom might enjoy being addressed by their middle initials as if they were entertainers. There was another John Woody who was John D. Woody's uncle, and the latter was likely named to honor the former. Uncle John Woody did not play a role in the Union Telephone story.

There are two individuals named James Woody who are important to this tale. James H. Woody, as is his preference, is referred to as "Jim," and James D. Woody is referred to by his full name. In addition, there are similar name combinations that might confuse the reader. Lottie M. Woody was founder John D. Woody's sister, while Lottiebelle Woody was his daughter. Bonnielee Woody is John D. Woody's daughter, and Bonnie Jean is his granddaughter. The family contains a Howard D. Woody and a Howard G. Woody. The former is referred to simply as Howard, and the latter as Gene.

This book is the narrative of how this family-owned company managed its century-long journey of creating a communications empire in one of the least populated portions of the nation. The Union Telephone Company has been on a wild rollercoaster ride, from barbed-wire telephone lines connecting a few hand-cranked magneto phones to a modern digital communications system connecting individuals around the globe via the latest technology.

A rodeo rider feels successful if he or she can stay on a bucking horse for eight seconds. Hang on to that bucking horse for a hundred years and you will understand the level of the accomplishment for Union Telephone Company in reaching this milestone. Over the course of a century, there were many junctures where the company could have ridden off into the corporate sunset by splitting up and selling its assets to competitors.

In 1948, after decades of running the company at a loss (and with Union's founder ready to retire), the family offered the company to Mountain Bell/AT&T for the bargain price of one dollar. The corporate giant's brusque refusal reminded the family how much they loved the company, and also why Union Telephone was founded in the first place. After that, there was no thought of selling the company. In its centennial year, Union Telephone's assets and stock value were in the tens of millions. Perhaps Mountain Bell/AT&T management learned from this episode to not look a gift horse in the mouth. Union Telephone, a family-owned and operated business, is poised to compete in one of the most cutthroat industries in the world for years to come. The company's ride is not over by a long shot.

The telling of this story relies heavily on dozens of interviews conducted with Union Telephone employees and members of the Woody family. Although this book is sponsored by the Union Telephone Company, it is not intended to serve as an advertising vehicle. The intent of this book is to blend the family stories and corporate history with the historical record. The rise of Union Telephone and the family who built it is a compelling tale. This book was written in the hope that the generations of the family yet to be born, the employees, the local community, Union's customers, and the general public will better understand the remarkable accomplishment of building a modern communications system in the Rocky Mountain West. Toward this end, John Brooks's centennial book about the industry, *Telephone: The First Hundred Years*, provided excellent source material about the larger story of changes in the greater phone industry, and it is recommended as a companion to this story.

Rural customers are well aware of changes in telephone services and technology. They demand service features that are available in the urban areas. In response to their customers, Union Telephone/Union Wireless often found itself on the leading edge of technological changes. Union customers have not been shy about demanding that their services keep up with those of urban America. Union Telephone became a leader in embracing new technologies and solving the issues of adapting new technologies to the brutal climate of America's Outback. That is remarkable when one considers the company is headquartered in a state that currently has a higher population of antelope than humans.

THE CHANGING FACE OF TELEPHONE SERVICE

Writing in the 1880s, Ambrose Bierce offered the following definition of *telephone* in his infamous Devil's Dictionary: "Telephone, n. An invention of the devil which abrogates some of the advantages of making a disagreeable person keep his distance." "Bitter" Bierce, as he was nicknamed, was speaking for the many people who have been electronically hounded and pestered by unwanted guests. The telephone has the advantage of allowing one to put off an unwanted visit by simply hanging up and terminating the call, rather than the more aggressive act of slamming a door in the disagreeable visitor's face. However, even Bitter Bierce was unable to fathom how people, disagreeable and otherwise, might use the new invention to close the distances between them.

What is it about telephone service that so captivates us? When the telephone was invented in 1876, it was considered a novelty and a toy. On the day that Alexander Graham Bell made his new invention functional, the headlines of newspapers across the nation announced that George Armstrong Custer and 267 troopers under his command had fallen on a barren battlefield near the Little Big Horn River. That historic day seemed destined to generate lasting icons. More than a century later, the telephone clearly has a greater effect upon the lives of billions than that famous battle. Custer's last stand has become the subject of more books and motion pictures than Bell and his invention, but being popular in books and movies does not necessarily reflect impact. George Armstrong Custer, Sitting Bull, and Crazy Horse made history, but Alexander Graham Bell changed the world.

At first, Bell's contraption scarcely seemed to be capable of becoming the engine of commerce and interpersonal bonding that it has grown into. Bell spent much of his early days in the telephone business defending his patents, as there were many who claimed to be the telephone's true inventor (Coe 1995). Electrical fluctuations in the early lines that Bell set up generated so much interference and random noise that many people believed the telephones were haunted and channeled the dead back into the world of the living. This was hardly an auspicious start for an industry that eventually grew to the status of a worldwide public utility, and is now considered almost as essential as electricity, water, trash removal, and sewer service. But that is how it was.

The telephone has been through almost unimaginable changes over the years. The restriction of being bound to wires attached to poles has been overcome, and telephone lines may soon be obsolete, replaced by an invisible and intangible network of radio signals that can even connect consumers flying far above the earth in airplanes. The heavy, wooden crank (magneto) phones were replaced by rotary-dial phones, which were then replaced by touch-tone phones. The single telephone positioned in a communal area of the home was replaced by multiple telephones throughout the building, often with a telephone handy in every room of a structure. More-private individual devices replaced communal telephones. Telephones connected to landlines were replaced by wireless telephones with base stations tethered to a landline, and eventually these became completely wireless telephones.

In the early days of telephone service, most people shared not only their telephone receivers, but they also shared their telephone lines with others. The quaint "party lines" of the past fell from grace, as consumers demanded private lines. However, something of the spirit of community was lost in the conversion from party lines to private lines. Telephone's age of innocent civility had passed.

So much has changed about phone ownership and access. The telephones in consumers' homes used to be owned by the telephone company and rented to the customer. Telephone booths were once such a familiar sight that it didn't stretch plausibility that the premier comic-book superhero of the day found these omnipresent chambers sufficiently available to change from his secret identity into the man of steel. When cellular networks and the ability to purchase "throw away" phones at most travel stops and department stores became commonly available, the telephone booth became both a dinosaur and an increasingly common display in museums.

Over time, the manner in which calls are connected has transformed from professional service to self-service, and lastly to a service where the device performs actions for you upon command. Phone connections, originally routed by operators, were changed to direct dialing, then direct touch-tone dialing, and then routed to Internet addresses by computer chips. The electronic voices of the computerized assistants found on some smartphones have become more recognizable than the voices of our politicians and celebrities.

Complete communications mobility has made quaint the vestiges of culture that once showcased the status symbolism of having a telephone in the home. Telephone rooms, telephone nooks, and telephone tables (once an integral part of many households) found their ways to museums and antique collections, moving from practicality to nostalgia. These rooms, nooks, and tables have gone the way of the nineteenth century "swooning couch," once a popular feature in Victorian homes. The status attached to telephone ownership has shifted to the technology contained within the communications device itself. This has created an industry propelled by the need to keep changing just as fast—or faster—than the competition, as well as by the desire of consumers for greater smartphone multi-functionality, such as image processing and computing capabilities. As with the fashion industry, change can almost be seasonal.

The telephone has morphed from a communication device to an entertainment center, concert hall, theater, library, comedy club, weather center, financial advisor/manager, telegraph, radio, camera, valet, doctor, psychiatrist, religious advisor, veterinarian, day planner, tracking device, advisor, casino, restaurant critic, answering service, guide, alarm clock, photograph album, compass, computer, and television, along with its occasional use as a device to talk directly with another person or their own electronic valet. Today, a person can contact another individual through their telephone number without either party using a telephonic device that would be recognizable to Mr. Bell.

The camera function of a mobile telephone brings the power and immediacy of imagery to the news stories of the day like no other period in history. Already, any place and any subject that can be observed can and will be recorded and distributed far and wide. Privacy has almost become an outdated concept.

The telephones of fiction have been far outstripped by the capabilities of the telephones of fact. The cross-pollination of the telephone system with computers and the Internet has turned the telephone into a handy gateway to the world. The flexibility and convenience of modern telephones has changed them into remote controls for so many devices in our world. One can use the telephone to turn on household systems and appliances, such as climate controls, sprinklers, pool cleaners, lighting, and security systems. A vehicle's driver can use the modern telephone to link into traffic cameras and get real-time information about road conditions. Smartphones can tell a user where they are when lost, and provide directions to just about any business or attraction.

As the complexity of communications devices has changed, the companies providing communications services have also been transformed. In the beginning, companies had to convince consumers that having a telephone was worth the expense. Because customers were tethered to landlines, they often spent decades with the same service provider, which fostered the "natural monopolies" held by telephone companies. There simply was no competition, and one's local service was part of the home team. As the regional and state monopolies on telephone service were broken up, consumers found themselves with real choices in telephone service.

As the ferocity of the competition increased (combined with consumers gaining unprecedented mobility), customer loyalty tended to fly out the window. Service providers fought tooth and nail to lure customers from their competitors, and they attempted to enforce a form of customer loyalty through service plans and contracts. Even so, it was difficult to keep customers in a market that had become so technologically driven. The consumers preferred the best deals above maintaining a long relationship with a more reliable service provider. It became a "What have you done for me lately?" world. In such an atmosphere, many service providers embraced the horror of a fickle customer base, placing less emphasis on customer service and more on promotion. Regarding this matter, the Union Telephone Company has maintained its traditional character of emphasizing the importance of customer service, and, in adopting that philosophy, has broken away from the herd. Union Telephone still functions very much like a local telephone company, even though it has become more of a regional entity, and it is making inroads on national and international markets.

The telephone has been both an instrument of social change and a victim of that same change. Few inventions require that society adapt its customs to serve the technology. The adoption of the telephone required its own social conventions. Alexander Graham Bell promoted "Ahoy" or "Hoy" as the appropriate greeting for use on his new invention. Much to his chagrin, the public chose "Hello." The invention opened up whole new careers for women as operators, or as they were nicknamed, "Hello Girls."

Social norms were established, and they dealt with the appropriate times to make or receive calls, party-line ethics, who could answer the phone, methods for ending calls politely, appropriate telephone behavior, proper uses of the telephone, and what constituted harassment. The kind of telephone, number of telephones, and use of long-distance services became status symbols.

Phones have managed to insert themselves into almost every aspect of life. To future historians, it may seem that the communications devices we use have domesticated the human race. Telephones have allowed us to act on our best and worst impulses. Bitter Bierce labeled the telephone as an invention of the Devil. But surely better communication between people across the globe deserves to wear a halo more than a set of horns.

Let us dial back into the past and begin our story. Our first connection takes us to Mountain View, Wyoming, one hundred years ago.

INTRODUCTION

Communication is probably one of the most important things we have. To me it was just a challenge.

—Howard D. Woody in a 2008 radio interview

MOUNTAIN VIEW, WYOMING, 1914: JOHN D. WOODY HAD A DREAM

In 1914, the final poles for the nation's first transcontinental phone lines were placed and strung with wire. Through the wonder of this evolving technology, people on the East Coast could converse with friends and family on the West Coast starting in 1915. The new long-distance lines compressed the vast distances separating Americans as never before. It was an exciting time to be in the telephone industry.

With Europe mobilizing its enormous armies at the start of the First World War, telephonic communication took on strategic importance. The ability for commanders (separated by miles from the frontline and no man's land) to connect directly with junior officers allowed the commanders to meddle in the tactics employed by the junior officers to an extent rarely attained in the past. Mr. Bell's 1876 invention was remaking both the nation and the world.

John D. Woody looked out at the verdant pastures of the peaceful Bridger Valley in southwestern Wyoming and saw opportunity. Wyoming had a sparse population, somewhere between 146,000 in 1910 and 194,000 in 1920. To place those figures in perspective, Denver, Colorado, had a population of 213,000 in 1910 and 256,000 in 1920. The major phone systems, in large part built by AT&T/Bell Telephone, were not serving the isolated ranches and small commu-

Union Telephone Company's founding father, John D. Woody.

nities of America's vast rural lands. John D. knew that a reliable, modern telephone system must improve the quality of life for his neighbors in the Bridger Valley and beyond. If the majors were not going to provide service in rural Wyoming, he was determined to take on that challenge.

The need for trustworthy modes of communication was made all the more crucial by the mammoth distances and harsh environments of the Rocky Mountain West. Wyoming families relied upon each other for survival. A grim truism of life in the Cowboy State was that neighbors had to be able to rely upon each other during emergencies. The bonds between neighbors were made strong by necessity. Success in raising livestock and crops was more likely when neighbors worked together. The great American Outback, of which Wyoming is but a small part, was a world populated by families of rugged individualists who, ironically, were bonded together by their isolation.

Enterprising rural communities and individual ranchers pulled up their sleeves and set to work erecting a patchwork of independent, local telephone systems. It was a typical Wyoming solution to a problem: don't wait for outsiders to show up and make life better—pitch in and take care of things yourself. A good example of this occurred in the Bridger Valley. When Union Telephone was founded in 1914, the community of Lyman was already operating an independent telephone system that was limited in scope to a few homes and businesses within that town. The Bridger Valley boasted several detached ranch lines as well. None of these independent systems were linked together, and they are best envisioned as little islands of telephone service. John D. Woody's dream was to link these separate telephone lines (many of which were merely lines stapled to fence posts) and provide modern switchboard service. In the early years of the twentieth century, the bringing together of detached parts and the joining of independent people into a common group was referred to as a union, hence the name for John D. Woody's new company, the Union Telephone Company.

Pulling together the hodgepodge of existing independent lines and exchanges that were widely dispersed throughout rural Wyoming was not an easy task. The effort required exceptional negotiating talents, patience, cooperative neighbors, luck, and the sheer strength of will and body. Patience as a virtue has been much underrated. In rural Wyoming, perseverance was a trait that often created success.

The Bridger Valley, while not the most hostile environment in Wyoming, was often a lineman's nightmare. Winter can rage like a beast, with howling winds approaching speeds of one hundred miles per hour. These winds turn the already cold, arid climate into something akin to nature's freeze-drier. Southern Wyoming wind chills all too frequently drop surface temperatures to tens of degrees below zero. Blizzards can pile up several feet of snow within the short space of a day or two. Wyoming's infamous winds made that difficult experience almost unbearable by rearranging the piles of fallen snow into a constantly shifting landscape of frozen drifts, with some as hard as stone and others as powdery as cake mix. Almost incessant ground blizzards can create whiteout conditions, when even determining up from down becomes difficult. Snow falls in the higher altitudes of Wyoming almost any month of the year, occasionally resulting in years that lack a real summer.

In many respects, the warmer months in America's Outback are even more difficult than winter. They're characterized by blinding dust storms, muddy roads, flash floods, tornados, pelting hail, lightning, brush fires, and seemingly constant winds so strong that opening two doors in a pickup truck at the same time is an act of courage. Rural Wyoming is no place for weaklings. It was in this environment that John D. Woody chose to follow his vision.

While John D. was the dreamer who conceived this unique blend of neighborhood service and a family company, his son Howard was the driving personality who molded the company into a serious competitor. The successive generations of the Woody family expanded the company into a regional power, and they're making strong inroads into the national and global markets. John D. laid the foundation, Howard built the superstructure, and the later generations took the company to the cutting edge of technology, reaching for horizons the founder never dreamt of.

One might say John D. had a knack for recognizing the opportunity for a telephone company, and that Howard was born to be in the telephone business. When Howard's mother delivered him in a room behind the Union Telephone switchboard (where she worked as an operator), destiny appeared to clearly be showing Howard the way to his future. Howard's providence continued to drive the company as it began its second century. Successive generations of the Woody family have taken up the dream and continue to lead the company into the twenty-first century.

This book is the story of the Woody family's first hundred years of—as they proudly say—"bringing neighbors together." This tale is a story of perseverance and ingenuity. Years of hard work, patience, and risk taking have paid off. Indeed, after surviving a century, few companies can unhesitatingly claim that their best years are ahead of them as they look forward to the challenges of a second century.

Our next connection takes us further back in time to a dismal, stormy night near Snyder, Texas, several years before the founding of the Union Telephone Company. On this night, the events were set into motion that would ultimately result in the founding of the company.

Chapter 1

Lone Star State Tribulations and Cowboy State Opportunities

A dream doesn't become reality through magic; it takes sweat, determination, and hard work.

—*Colin Powell*

NEAR SNYDER, TEXAS, CIRCA 1895

The cattle milled about restlessly, and the moon had barely risen when a bank of heavy, gray clouds drifted across the shimmering disc, seeming to swallow it whole. The cattle sensed the change in the wind and smelled that dry, earthy scent that meant only one thing: a big storm was coming. The local Texans called these storms Northers, as the worst of them came out of the north. Being caught out in the open during a Norther was a memorable experience, and one a person might not wish to repeat. The cold winds of such a tempest whipped a tidal wave of dust, rain, and sometimes sleet or snow across the open tabletop of the Llano Estacado, a Spanish name often translated as palisaded plain. However, the locals more correctly interpreted the name as the staked plains or staked level.

Feeling the atmospheric changes, the apprehensive cattle began to low. They wandered about nervously as the first fingers of the storm front began to assault them with clouds of choking dust, followed by stinging rain and hail. Soon the skies flashed with lightning, and deafening peals of thunder echoed in the clouds. Terrified, the cattle looked for a way to relieve their discomfort, but there was no shelter on the open prairie. They were at the mercy of the storm, and Mother Nature was not in a forgiving mood that night.

The bovines' best hope to escape the discomfort of the storm lay in movement. Instinctively, the herd began to crowd closer together. A few cows on the outer fringes of the mixing mass of frightened animals began to lope away from the oncoming storm in the vain hope of outrunning the tormenting torrent bombarding them. These fleeing cattle were shortly joined by a few more, and soon all were following, jogging along after the leaders.

Before the herd began its exodus, the cowboys sensed a change in their livestock and sprung to action to try and head off the spreading panic. No one could stop what was about to happen. There weren't any fences or box canyons to help the wranglers corral the restive animals. The trickle of escaping livestock grew, and in short order they moved as a wave—inexorable, unstoppable, and inevitable. They passed right by the cowboys, who flailed their arms, whistled, and impotently shouted in the vain hope of turning back the escaping cattle.

As the storm intensified, frightened cattle started trotting *en masse,* leaving the outriders far behind. The cutting dust and sleet pelted the struggling riders, and it was not long before the cowboys struggled to keep the herd in sight. The outriders did their best to follow the sounds of the herd, but the pelting rain hammered their hats and gear so loudly that the riders could hear little else. Blinded and deafened in the maelstrom, they lost contact with the herd when it turned in a new direction. The horses became more difficult to handle, and their riders endeavored to remain mounted. The realization that they too were in trouble whipped them almost as hard as the Norther's torrents of rain.

As the storm intensified, the jaunting lope of the cows accelerated into a panicked stampede. The herd turned to run in a direction opposite the stinging rains. The cattle in the front may have only had a second or two to realize that their flight had led them to the edge of a hundred-foot-high precipice, one of the Llano's many palisades. The startled lead animals attempted to plant their feet deeply in the prairie soil, but it was too late. The press from the scores of frightened beasts behind them propelled the leaders over the cliff face. Inexorably, almost the entire herd plummeted into the abyss, unaware that their flight had propelled them towards danger, and not away from it.

Through the howling winds, the sounds of screaming cattle and the crashing of heavy bodies against the flinty cliff's base filled the night. Bones splintered like matchsticks, and blood mixed with the rain that ran down the cliff face. Animal after animal blindly tumbled into the chasm. They piled atop the dead and injured animals below, until so many had fallen that their bodies formed a ramp. The last of the stampeding cattle managed to stumble down a precarious path to the bottom of the cliff, treading on the dead and writhing bodies of the less fortunate members of the herd. The few fortunate surviving cattle disappeared into the dark, leaving the outriders to deal with the aftermath.

LIFE ON THE LLANO ESTACADO

No one ever said that life on the Llano Estacado was easy. The semi-arid environment of one of North America's largest tablelands marks the southern limits of the Great Plains. The Llano has cap rock escarpments (some of them hundreds of feet high), which are found mainly around its edges. These escarpments create the palisades (or stakes) alluded to in the area's name.

Once, rivers of buffalo navigated these plains. The patchwork of tall and short prairie grasses creates an ocean of undulating movement. By the year 1900, the wooly giants had been hunted to the brink of extinction for their meat, tongues, and fur. Throughout the Great Plains, millions of buffalo skeletons were collected and ground into fertilizer. The reduction of the buffalo herds opened up vast expanses of grazing land for cattle and other non-native livestock.

In 1892, Snyder, Texas, was a community southeast of Lubbock with a population of roughly six hundred souls. The town had colorful sobriquets such as "Hide Town" and "Robbers' Roost." The Snyder brothers' cattle company grazed the Llano beginning in the late 1860s, and by 1895 they had established the Texas Livestock Association (Hope 2011). Snyder, Texas, was cattle country, pure and simple.

The Llano was one of those places that tested all of its inhabitants. The staked plains of Texas were the battleground for a series of wars that swept through the area beginning in the late seventeenth century, only concluding with the surrender of Quanah Parker and the last of his band in 1875. Prior to Quanah Parker's surrender, the staked plains were within Comancharia, one of the largest Native American empires in history. The Comanche had fought the

Azariah B. Woody's early life in Texas reads much like a dime-store novel. Among the first homesteaders in Scurry County, he farmed and worked as a blacksmith prior to becoming a stock grower. There were many adventures during those days, including numerous buffalo hunts and several skirmishes with hostile Native Americans.

Kiowa, Osage, Lakota, Pawnee, Arapaho, Cheyenne, Ute, Apache, Spanish, Mexicans, Texicans (Republic of Texas), and Americans (Confederate States of America, and United States of America) for control of this vast area. For centuries, the Comanche were capable of putting thousands of formidable warriors in the field, and they occasionally raided as far as a thousand miles outside of their expansive homelands. It took their enemies hundreds of years to chip away at the Comanche's power, but in the end the Comanche nation was overwhelmed by a relentless tide of expansion and settlement.

The history of the Llano reads like a Who's Who of the West, including notables such as Buffalo Hump, Quanah Parker, Cynthia Anne Parker, Francisco Vasquez de Coronado, Juan de Oñate, Sam Houston, Nelson Miles, Ranald McKenzie, Kit Carson, Earl Van Dorn, Robert E. Lee, and William Tecumseh Sherman, all of whom had their parts to play in the Llano's story. The Republic of Texas created their most famous law enforcement branch, the Texas Rangers, to help stem the Comanche tide. The Rangers' daring use of guerilla "hit-and-run" tactics effectively kept the Comanche off balance and opened the Llano to ranching.

In the mid-1890s, the family of Azariah Brown Woody (sometimes called A. B.) was among the many newcomers on the Llano. Born in Missouri in 1852 to teamster Robert B. Woody and his wife Matilda Green Woody, Azariah Brown Woody matured into a tall man who was reputed to be six and a half feet tall. As was common for the inhabitants of Newton County, Missouri, he learned to handle livestock at an early age. A. B. was the fourth child born to Robert and Matilda, and he was their second-oldest son. Azariah and his brothers (George, John, Edmond, and Dave) and sisters (Elizabeth, Serilda, Mary, and Eliza) grew up in the shadow of a nation at war. Two battles of the Civil War were fought in their home county, both of them at Newtonia (September 30, 1862, and October 28, 1864). There were significant smaller-scale hostilities throughout Missouri over the course of the war, and the Woody family could not have helped being affected by the conflict.[1] In 1878, A. B. married Isabell Hazelwood,[2] a French-Creole girl from Louisiana. Isabell was eighteen years of age when she took her marriage vows, and she may have been showing the first signs of tuberculosis at that young age. Together, Azariah and Isabell raised four children: Matilda (born 1879), John D. (born 1883), Lona Isabell (born 1887), and Lottie (born 1891).

Isabell Hazelwood Woody and Azariah B. Woody on their wedding day.

During the late 1870s, A. B. and his family lived near Cedar, Missouri, in Callaway County. Azariah and his younger brother Dave farmed a spread there. Matilda was born during this period. We know little about this period in A. B.'s life, except that he and other family members were soon farming in Johnson County, Texas.

[1] *Records from the Civil War indicate that a Robert H. Woody served in Company G of the Tenth Missouri Infantry Regiment. Given the similarities between the letters B and H when written in script, it is remotely possible that Azariah's father served in the military during the war. Both the Confederacy and the Union fielded units designated as the Tenth Missouri Infantry. The Federal unit appears to have been recruited from northeastern Missouri counties, while the Confederate unit was recruited from central Missouri. It is unlikely that Robert B. Woody, living in southwestern Missouri, was mustered into either unit.*

[2] *The US Census spells her name Isabel and Isable. The spelling used here conforms to her tombstone.*

In 1880, both Robert and Matilda died. The 1880 US Census places Azariah, Isabell, Matilda, and Dave Woody living on the staked plains. Isabell's occupation listed in the census of that year indicates that she kept house. Shortly after the collection of the 1880 census, the Woody family moved to a ranch near Snyder, Texas, in Scurry County.[3] Once in Scurry County, Dave no longer seemed to have been a part of Azariah and Isabell's household.

Azariah was a hard worker and easily adapted from farming to ranching. He loved raising cattle. A. B. was just the kind of man one expected to find out in the staked plains tending his herds of lowing cattle: a tall, lanky man sitting comfortably in the saddle. His weather-beaten skin conveyed the years of lonely work spent constantly working outside in good weather and bad.

Azariah had a strict code of discipline and was determined to raise children who were tough enough to take on the challenges of ranch life. If he were a character in a Western novel, Azariah might be introduced as a man who was unbending, and as flinty as the land he ranched. Pictures of Azariah reveal a thin man with a dark complexion and narrow, intense eyes. The slight down-turn at each of the corners of his mouth gives him a stern expression. The images reveal a serious man in charge of his own fate, and one who was not to be trifled with. It is not difficult to imagine such a man sitting tall in the saddle and swinging a stout rope as he cut cattle out of the herd for branding.

Azariah attempted to breed hardy varieties of Hereford cattle. Though by nature resilient animals, the Herefords needed to adapt to a climate that varied from unbearably hot and humid to icy, cyclonic Northers. Despite the abundance of grasses, success was not guaranteed to those who wished to harvest the bounty of the Llano. Sufficient water for raising cattle was a scarce commodity throughout much of the staked plains. The dreaded Comanche were no longer an issue, but Mother Nature still had plenty of fight in her, and she did her best to remind those who lived on the staked plains that she was calling the tunes.

Azariah raised the cattle with the help of his family. John D. Woody, as the only male child, found himself with substantial duties related to tending the

[3] *Other Woody family members apparently moved to the area around Snyder as well. John Woody (A. B.'s brother) is recorded as passing away in Snyder in 1934.*

pride of the ranch, the Herefords. He likely shared these duties with his father and his uncle. It was a hardscrabble existence, and to make ends meet, Azariah was forced to spend much of his time drilling water wells and pumping livestock water for other ranchers. This left his son John D. (thought to be twelve years old at the time) the task of watching the livestock for an entire summer while his father was away earning wages; his uncle was likewise gone. This was a job for which Azariah promised to pay John D. with a brand new saddle. This offer of a new saddle motivated John D. to do his best, and it got him through the tedious months of toil. It was a job of great responsibility for such a young man. When the summer passed and payment was made, John's heart sank as his father produced a used and beat-up old saddle instead of the new saddle, the promise of which had been the impetus for the hard work John D. had performed that summer. It was a shattering disappointment to John D.

The uncompromising Azariah wanted to use the promise of the new saddle to teach young John D. a life lesson about expectations versus realities. The boy protested getting a raw deal, and Azariah is said to have replied: "Let that be a lesson to you. Life is hard. Promises don't mean anything." It was a tough lesson for a young boy, but the Llano had a way of handing out hard examples such as this. At that time, John D. learned a lot about patience and how to deal with disappointment. These were virtues that would help get him through the tough times to come.

The Woody family eked out only a modest living on the broad grasslands of the Llano, but they were succeeding. The herd was increasing, and the children were growing up. It is not certain when the family felt the hammer fall on them in the form of the Norther (described at the start of this chapter) that all but wiped out their herd of cattle. Family members suggest that it may have been as early as 1895, but surely occurred between that date and 1900.[4]

It took the disastrous loss of cattle from that deadly Norther to force Azariah to reconsider his choices. For A. B. and his immediate family, the losses incurred during that event were devastating. In the aftermath of the incident, the coming

[4] *Census records show the Woody family still living in Snyder in the year 1900, but the family maintains that Azariah and his family were already in Wyoming before that date. The family history suggests an arrival in Wyoming between 1895 and 1900.*

winter took on a grim and deadly aspect. How was the Woody family to survive a harsh Llano Estacado winter with no livestock other than a few horses that had been spared from the catastrophe? No one ever said that life on the Llano was easy, but this was too much to bear. Azariah had difficult decisions to make.

In 1895, thousands of hopeful prospectors were already on the move to the Klondike gold rush. However, the northern goldfields were no place to raise a family. Azariah heard that there was work in Wyoming on the Oregon Short Line, a railway between Granger, Wyoming, and Huntington, Oregon. Wyoming was cattle country, and there were good prospects that the wages earned while working for the railroad might be parlayed into a new ranch and a new herd of cattle. At the same time A. B. and his family were considering emigrating from Texas, the Oregon Short Line was hiring teamsters and laborers. A. B. decided to take his chances and trade the staked plains of Texas for the high-altitude, intermountain basins of Wyoming. It was a big risk, but the Woody family was up to the task.

In the late 1890s, the Oregon Short Line was in the process of being converted to standard gage rails. The railroad had started as a Union Pacific Railroad subsidiary in 1881, but became a separate entity when the Union Pacific Railroad went into receivership during the Panic of 1893. For a time, the railway operated independently as the Oregon Short Line and Northern Utah Railway. However, the reconstituted and reformed Union Pacific Railroad reacquired the route in 1898.

The Woody family packed their possessions into a wagon and began their long trek to the Cowboy State. There were many routes the family could have traveled on their journey of roughly one thousand miles. Tried and true trails, such as the Cherokee, Chisholm, and Overland Trails, offered possible avenues. Additionally there were numerous freight roads, stage routes, expansion roads, military roads, commercial paths, and railway corridors that the family could follow in the late 1890s through the early 1900s.

In the aftermath of the devastating Norther, the family still owned good teams of horses. Azariah decided to take those teams and follow the roads through Amarillo to Denver. From the Mile-High City, the Woodys continued up the front range of the Rockies until they struck the combined route of the Overland Trail and Cherokee Trail that ran through Bridger Pass.

In Denver, the family may have had a photograph taken. Howard Woody remembered the photograph as depicting the wagon, the team, and the family members, but the image seems to have been misplaced and was not available for the publication of this centennial remembrance.

The Woody family's chosen route to the terminus of the Oregon Short Line at Granger, Wyoming, cuts a more direct path from the front range of the Rockies to Fort Bridger and the Bridger Valley. In the late 1890s, the Overland and Cherokee trails intersected the Union Pacific Railroad near Point of Rocks, Wyoming. The Bridger Pass alternative was well worn by the passage of thousands of emigrants bound to the California gold fields in the 1850s, and later by the passage of freight, stagecoaches, and people traveling along the relatively better defended Overland Mail route.

The main advantage of traveling by way of Bridger Pass was that the gap through the foothills trimmed several days off of the journey. The downside to this option was the ruggedness of the terrain encountered, combined with what must have been some of the poorest drinking water between Texas and Wyoming. Much of the Bridger Pass route followed the aptly named Bitter Creek, and includes oases of dubious quality, such as Sulfur Springs. At Granger, the family reached the start of the Oregon Short Line, where Azariah sought employment.

Travel by wagon was a grueling process. Most chose to ride or walk alongside the wagons, rather than be pummeled hour after hour as the poorly sprung (or often un-sprung) wagons of the time period bounced over unimproved roads. Unlike modern automobiles, wagons in the late nineteenth century usually lacked suspension systems that would have absorbed or reduced the shocks created by travel. Riding in a wagon was much like having one's body pummeled with a club. Each bump along the unimproved roads sent bone-jarring shockwaves through the wagon. Stagecoaches had more robust and better-designed leaf springs for the comfort of the passengers. Even so, a frequent complaint of the relatively pampered stagecoach customers was how unendurably rough the ride felt.

Unless the Woody family had brought a significant amount of beasts of burden to haul their wagon(s), they probably averaged approximately fifteen miles per day. A supply of fresh livestock and replacements that could be exchanged for tired animals might improve the overall speed of travel, but not by much.

There is no evidence that the Woody family had any teams of oxen to aid their horse teams. Horses generally had less lift capacity to haul heavy loads, but tended to walk faster than the plodding—but stronger—oxen. The oxen also were superior in their ability to subsist on native vegetation that was available along the trail routes. Horses are more fragile than the sturdy oxen, which can eat native plants that might sicken horses.

The days on the road likely started early. They would break camp at first light, and then at mid-day they would take a long break (often called a "nooning") to rest the animals and allow them to graze. The family likely established a camp in the afternoon (if possible), and set up their shelters for the night. The animals were tended, led to water, and allowed to graze. Equipment was maintained and repaired in camp. The animals were guarded during the night so that they would not wander away, leaving the travelers stranded. It was crushingly hard work just to get through all the chores of the day, to which was added the effort required to walk or ride through the backbreaking terrain for miles on end.

This cycle of work and walking was repeated, good weather or bad, for roughly ten weeks, the amount of time needed to travel from the Llano to the Bridger Valley at fifteen miles per day. The trip might have taken longer if the family rested on Sundays or were held up by illnesses or major repairs. The long journey most likely was completed before three months had passed.

A NEW HOME IN THE COWBOY STATE

In 1890, the Territory of Wyoming was admitted to the United States as the forty-forth state. The transition from territory to statehood had been a bumpy adventure. Throughout the previous decades—and prior to it being established as Wyoming Territory in 1868—portions of Wyoming had been within several other territories, including Oregon, Idaho, Utah, Washington, Dakota, Nebraska, and the Republic of Texas. The new territory was formed in the same year that the Fort Laramie Treaty ended Red Cloud's War. The discovery of the Miner's Delight deposits in the South Pass area in 1867 set off a short—but intensive—gold rush in Wyoming.

For decades, Wyoming had been America's way station to other destinations. The area was crossed by several of the nation's greatest wagon roads, including the Oregon, California, Mormon Pioneer, Cherokee, and Overland Trails.

Together, these trails—along with the numerous short cuts, cutoffs, and alternative routes—created a spiderweb of over a thousand miles of wagon roads throughout Wyoming. Ranch roads, haul roads, stage roads, livestock drive routes, railroads, and military roads expanded from the system of wagon trails. The Union Pacific Railroad corridor coincided, in large part, with the wagon roads, deviating mainly to incorporate important coal resources into the railroad's route. Wagon trains were fueled with grass, but railroad trains initially ran on wood, and then later, coal. For both kinds of trains, the availability of water was a must. The availability for fuel and water established the routes from which Wyoming's economy grew.

The Pony Express, the Overland Mail, the Union Pacific Railroad, and the intercontinental telegraph made Wyoming a keystone in maintaining the national communications system. The Pony Express became an iconic image of the West. However, horseback was an inefficient and costly manner in which to carry mail between St. Joseph and Sacramento. More-efficient semaphore systems had been used in Europe since the early 1790s. These systems relied upon tall towers that were set within eyesight of one another, atop of which people used lights or large mechanical arms to signal messages to the next tower. This system of towers is called a semaphore line. It was developed by the Chappe brothers, and promoted by rulers such as Napoleon Bonaparte (Standage 1998, 6–14). With such a system, Napoleon, while in Moscow, was able to read messages initially sent just a few days earlier from Paris.

The sheer style and panache of the Pony Express captured the American imagination. The images of wiry young men spurring their horses along through the dangers of the Great American Desert were far more compelling than those of technicians positioning the arms on the windmill-like semaphore towers, or flashing heliographs (mirrors) to send coded messages to a another group of similar technicians at the next station. The Pony Express lasted a little less than two years (1860-1861), when the completion of the transcontinental telegraph—combined with the building of an overland mail route—made the system obsolete.

The Overland Stage Line was built almost hand in hand with the transcontinental telegraph system. Employing stagecoaches for carrying mail was slower than the Pony Express, but it allowed for the shipping of freight and larger parcels that were impossible for the swift ponies to transport. However, it was the telegraph that far outperformed the exotic system of relaying small mail pouches between galloping riders.

With the advent of the telegraph, it was possible to transmit messages across the continent in a few hours. The trans-Atlantic telegraph cables allowed equally rapid communication with Europe. In 1858, financier Cyrus W. Field helped bankroll the laying of the first trans-Atlantic telegraph cable. The project was heavily subsidized by the governments of the United States and Great Britain, both of whose naval vessels carried the heavy cable (which weighed about a ton per mile of cable) to the linkage point at the center of the Atlantic. Almost immediately, the 2,500-mile-long, half-inch-thick cable became a huge success, and it was heralded as a technological miracle. The euphoria lasted for roughly a month, at which point the cable failed. A second cable—laid in 1865—also failed, and was considered unfixable. A third cable was laid in 1866. A month after the completion of this third cable, the second cable was repaired, giving the world two working trans-Atlantic cables. Combined with an 1864 cable—built between Britain and India via the Persian Gulf—much of the globe was now within the reach of telegraphic communication (Standage 1998, 74-91).

The telegraph was capable of more than just handling simple messages in code. Rudimentary copy transmission machines included Alexander Bain's 1843 fax machine, F. C. Blakewell's 1850 copying telegraph, Giovanni Caselli's 1860 pantelegraph, and Ernest Hummel's telediagraph (Bellis, n.d.). By 1870, the technology had developed sufficiently that French commanders were able to receive telegraphic situation maps from the battlefields of the Franco-Prussian War. The information superhighway was taking shape.

"LIKE NO PLACE ON EARTH"[5]

The Woody family was part of a new breed in the Cowboy State: people who emigrated from afar to settle Wyoming's intermountain basins. During Wyoming's territorial years, those who stayed in the state (and who were not Native Americans) tended to be trappers, traders, soldiers, freight or mail

[5] *This was a popular slogan of Wyoming's office of State Travel and Tourism, but to some people it suggested a hostile, forbidding, and alien landscape. It was replaced by the equally evocative, "Find Yourself in Wyoming."*

company personnel, railroad workers, or those providing services to any of these trades. These few souls were supplemented by thousands of immigrants brought in by the Church of Jesus Christ of Latter-day Saints (also known as the Mormon Church) to settle in "Zion." While most of the faithful settled in and around the Valley of the Great Salt Lake, many Mormons settled in the Bridger Valley.

By deciding to homestead in the Cowboy State, the Woody family was part of a small group of truly adventurous people. Many of Wyoming's newcomers during the nineteenth century had traveled to the state for temporary work, but ultimately were on their way somewhere else. Wyoming was often a speed bump for people traveling by land between the East and West Coasts. It was a great expanse replete with howling winds, a sea of sagebrush, wild-life, wilder inhabitants, violent storms, biting bugs, dangerous rivers, and lawlessness. Wyoming was a part of the Great American Desert, later called America's Outback.

The standard the Census Bureau set for labeling an area a "frontier" was that the area supported a population density of less than two inhabitants per square mile. By this measure, Wyoming was a frontier, and remained one well into the twentieth century. Azariah Woody saw opportunity in putting down roots in such a place. His wages from working for the railroad built the capital to establish the family as ranchers within Wyoming's fertile intermountain basins. The staked plains had toughened the family, and they needed every bit of that toughness to establish their new home in the Cowboy State.

Once Wyoming's fledgling livestock and mineral industries grew, the notion of settlement far removed from the transportation corridors became a viable option. The livestock industry that the Woody family—and other ranchers like them—built early on in Wyoming's history influenced and gave rise to the state's modern "cowboy culture." As compared to minerals, agriculture contributed only a tiny proportion to the state's economy, but it was the lifestyle of the rancher that Wyoming identified with. Even in Union Telephone's centennial year, symbols of Wyoming's cowboy roots were found everywhere, from automobile license plates to the name of the University of Wyoming's athletic teams: Cowboys and Cowgirls. The official state sport was rodeo, and many small communities built and maintained rodeo arenas instead of football fields.

Communities first sprang up around military posts, trading posts, stagecoach stations, mines, and railroad stations. The Bridger Valley was settled, in large part, due to the presence of a trading post (which later became a military post) that was founded by Jim Bridger and Louis Vasquez, called Fort Bridger. The Union Pacific Railroad completed laying tracks across the state in 1868, which improved communication with the remainder of the nation and allowed Wyoming ranchers to transport livestock to distant markets without undertaking the iconic, long-distance livestock "drives" popular in literature and film.

During the laying of the tracks, coalmining towns (such as Rock Springs) and "Hell on Wheels" communities (such as Bryan) sprang up in the shadow of the economic possibilities provided by the rails (Gardner & Flores 1989). The completion of the Union Pacific Railroad opened up new markets, and cattle soon began flowing up trails (such as the Chisholm Trail) from Texas and the Southwest. The expansion of the rail system (like the Oregon Short Line) created employment opportunities and ultimately helped convince the Woody family to settle in the Cowboy State.

As the Wyoming livestock industry flourished, the intermountain basins were all but cleared of buffalo. Bison held on into the middle of the twentieth century, with tiny relict populations in isolated places. Prior to the coming of the railroad, the cattle industry served the local military posts and other small communities. By 1890, large numbers of sheep had been added to the mix. Before long, the open ranges of Wyoming teemed with millions of sheep and cattle. In 1898, the official figures for Wyoming livestock were 706,000 cattle and 1,940,000 sheep. By 1914, the number of cattle had dropped to 583,000, while sheep boomed to 3,827,000 (Larson 1978, 367).

In a state where the land is almost exclusively a mile or more above sea level, there were few nineteenth-century crops hardy enough to grow. The success of crop production depended mostly upon the happy coincidence of sufficient rain, an adequate number of days without killing frosts, lack of prairie fires, lack of locust infestation, and a lack of devastating hailstorms. Dry farming in much of the state was doomed to failure. Development of agriculturally sound irrigation systems was slow. Crude irrigation systems often created additional problems for the long-term health of the cropland because of the salinization of the soil. Despite these issues, Wyoming became home to rugged settlers—like the Woody family—who were looking to carve out an existence in its wide-open spaces. By the time Wyoming came to statehood, there was a flourishing cattle industry, and the foundation of its mining industry was well established. Despite abundant grass, coal, and other minerals, there were growing pains ahead for the forty-fourth state.

Azariah and John D. took jobs on the Oregon Short Line. When they acquired their ranchland in the Bridger Valley, the Woody family settled roughly twenty-seven miles "as the crow flies" (but more like thirty-one miles on the existing trails) from the terminal point of the Oregon Short Line. Following the well-worn Oregon and California Trails from Mountain View to Granger, the trip took roughly two days by wagon, and that was when the roads were free of snow or mud. This traveling time could be shortened if a Union Pacific Railroad train was hailed and the conductor convinced to stop and pick up passengers. It is not clear if the train was a viable option for regular commuting to and from work. In any case, John D. and Azariah, by necessity, spent significant time away from their ranch, likely living near their work sites along the short line. The ladies of the family provided the bulk of the labor required to operate the ranch.

The division-of-labor pattern followed by the Woody family was a common one throughout the region. Multi-generational families often settled the rural areas and pooled labor and resources in a communal effort (Garceau 1997, 48ff.). Men of suitable age and good health often sought wage labor in distant industries, leaving the operations of the ranch or farm to the women, the elderly, the young, and the infirm.

John D. was still quite young—between thirteen and seventeen—when the Woody family reached the Bridger Valley. Because there were no child labor laws in Wyoming at that time, he was allowed to drive the family's teams of horses to scrape and smooth the railroad grade. At the time, the railroad was working on the replacement of narrow track with standard gage. In the early twentieth century, the Oregon Short Line purchased several small rail lines and began the moderately ambitious building program. This is possibly the reason workers such as Azariah and John D. found employment. These distant jobs might have kept A. B. and John D. hundreds of miles from their ranch for weeks at a time.

It is not clear how much Azariah and John D. were paid for their labor. Prior to its reorganization, the Union Pacific Railroad kept the wages of laborers low by employing a large number of cheaper foreign workers such as those from China, Ireland, and Japan. When the Union Pacific Railroad was put into receivership and reorganized, it helped facilitate the Panic of 1893. It is likely that the Woodys received slightly higher pay from Union Pacific for providing the family's teams for the grading work, as well as in recognition of the skills necessary to operate the teams. A reasonable estimate is that they each received a salary of between $600 and $1,200 per year, with the lower end of the pay scale being more likely. According to family sources, Azariah and John D. managed to earn sufficient funds to afford paying cash for the full asking price of Joe Hatch's two-hundred-acre Bridger Valley ranch in 1899 or 1900.

The length of time that the Woodys worked for the railroad before focusing on their ranch is not certain. The railroad was expanding well into the twentieth century, and was rebuilding track as late as 1902. If Azariah and John D. sought to be steady Union Pacific Railroad employees, a move to Green River (roughly thirty-two miles by wagon) or Evanston (roughly sixty-eight miles by wagon) was in order. Permanent work crews were stationed in both cities, and the railroad constantly repaired and improved vast stretches of track. Early in the twentieth century, Azariah and John D. ended their employment with the Union Pacific Railroad and focused on the ranch in the Bridger Valley.

LIVING NEAR THE OUTLAW TRAIL

The Wyoming that the Woody family settled in was in many aspects the archetypal Wild West. Butch Cassidy had been released from the penitentiary in 1896. He and The Hole-in-the-Wall Gang (which included Kid Curry, Black Jack Ketchum, and the Sundance Kid) preyed upon the railroad system, and then melted into the vast emptiness of the canyons and basins beyond the railroad corridors. Rustlers, robbers, murderers, and thieves such as Bill McCoy, Anne Bassett, Albert Bothwell, Tom Horn, Kinch McKinney, Bill Carlisle, and Cherokee Bill Pigeon were among the many outlaws who haunted the backcountry in and around southern Wyoming. While the presence of so many criminals and cutthroats makes for colorful history, it was a trying time for the Woody family to attempt to settle in such a place.

Reminders of Wyoming's outlaw past still dot the landscape. Bill Pigeon's grave, near Dutch John, is located in the heart of the Union Telephone service area. *Photograph courtesy of Terry Del Bene*

The outlaws infesting the West during this period were likely a matter of great discussion in the local community. The outlaw Bill Pigeon was killed in December of 1898 in a shootout with Ike Lee. Bill was buried in a canyon near the Wyoming border that now bears his name. This canyon is not far from Dutch John, UT, within an area in which the Woody family later established telephone service. Bill has not been allowed to rest in peace. According to local legend, his grave was opened and his skull was stolen (Tittsworth 1927, 196). Bill's purloined head created a cottage industry of local speculation as to where the errant skull wound up, with at least five completely different tales in circulation at any one time. Examples included Bill's head being moved to the grave of another outlaw tough guy whose head was collected as a trophy, and Bill's skull being kept at a sheepherder's cabin for use as a macabre puppet to entertain the bored pastoralists. The variety and detail of stories in circulation indicate that inhabitants in the area were enamored of speculating about the disposition of Bill's head—a matter that in Union Telephone's centennial year was still not resolved.

Wyomingites of the late nineteenth century loved to spin yarns about their outlaw neighbors, perhaps as much as trying to predict the fickle changes in the weather. As newcomers to the area, the lurid tales involving murder, trophy-head collecting, cannibalism, and grave robbery were likely unsettling. Wyoming probably took a little getting used to.

While the days of high adventure were in the recent past for the staked plains of Texas, the Woodys now found themselves in a place where the drama was all too current and all too close to home. Outlaws plied their trade wherever they believed money or valuable property was theirs for the taking. Even the Oregon Short Line, where Azariah and John D. took employment, was not immune to predations by criminals. On June 28, 1910, the front page of the Washington Post carried the following banner: "Robbers Loot Train." The story referred to a robbery by three masked men who held up the Oregon Short Line and took money from a hundred passengers. In the Woody family's new homeland, train robbers were the superstars of the outlaws.

Bill Carlisle, one of the last of the train robbers of the American West, continued to ply his trade in the region between 1916 and 1919 (Moulton 1995, 226-227). Carlisle was nicknamed the gentleman bandit because he did not steal the jewelry of the fairer sex, preferring to purloin the watches and wallets of the men. When Carlisle heard that innocent men were being held for one of his heists, he wrote a letter to a newspaper exonerating the prisoners. The gentleman bandit was apprehended a month after saving those innocent men. However, Bill escaped the penitentiary and was only returned to the penal system after being shot in the hand during his next train-robbery attempt. After serving out his time, Carlisle lived in Kemmerer, Wyoming, and worked in a hotel on the town triangle where the J. C. Penny home store is still in operation. It is likely that several members of the Woody family met the gentleman bandit, as he became a local celebrity.

HOMESTEADING

The Bridger Valley is within the "checkerboard," an area forty miles wide and centered on the Union Pacific Railroad line. The Pacific Railways Acts of 1862 and 1864 authorized the government to give alternate sections (each roughly one square mile) to the railroad company for twenty miles on either side of their main lines. This created an enormous windfall for the railroads. The granting of these large tracts of slightly contiguous land made it difficult to manage the open range because it created a complex land ownership situation where the railroad's sections of land touched only at the corners. Fencing a single section (one square mile, or 640 acres) required a minimum of four miles of fence wire per strand (and twelve miles of barbed wire for a three-strand fence) and approximately 850 fence posts. With millions of acres of rangeland involved, along with the need for regular fence maintenance, controlling one's range with fences was both cost and time prohibitive. The occupants of the checkerboard had to use the range cooperatively. Sometimes they did, and sometimes they were less than successful.

Cattle roamed between Union Pacific Railroad's land, federal land, the State of Wyoming's land, and privately owned land with impunity. A cow could simultaneously roam land and eat grass owned by several different owners. Keeping the livestock within an unfenced range was difficult at best. The difficulty of identifying the owners of the many maverick calves that were rounded up each year when herds became mixed complicated matters. It is little wonder that so many stock detectives patrolled the American Outback.

Once settled in their new home, Azariah and his family re-established their cattle herd. In order to do so, they adapted to the local methods for raising livestock. A semi-annual roundup system was in play. In the spring, cattle were branded around calving time. In the fall, fattened animals would be rounded up and shipped off to markets. This required infrastructure, corrals, and loading facilities that had not been common on the staked plains of the Llano. Cattle were often allowed to freely roam the open range in the winter months in Wyoming, which created issues of ownership for unbranded (maverick) calves. The other animals necessary for keeping the ranch operational required shelter against the harsh climate. Barns, chicken coops, hog pens, sheds, wells, outhouses, and fences were common features of homesteads during this era.

Joe Hatch's ranch already had a small house on the property when Azariah and John D. purchased the land. Most ranch houses of this time period were log or plank structures that were heated with a stove. Compared with modern dwellings, ranch houses tended to be drafty. When it was available, coal was preferred for heating; otherwise, the cutting, processing, and stockpiling of

Ranching in Wyoming often involved spending days on the range with the livestock. This image showcases a row of "sheep wagons" preparing to pull out. A wagon similar to these likely transported the Woody family from Texas to Wyoming.

large quantities of wood was necessary. The ranch houses were usually single-story affairs with only a few rooms. Family members often shared bunks during the winters to survive the frigid evenings. Oil and/or kerosene lamps illuminated the house. Interior plumbing and sewage removal were luxuries in rural Wyoming at that time. An electrical power system did not reach the Bridger Valley until 1940.

On December 1, 1911, Elinore Pruitt (a recent arrival to Burntfork, Wyoming) described a similar home from the period as follows:

> Every log in my house is as straight as a pine can grow. Each room has a window and a door on the east side, and the south room has two windows on the south with a space between for my heater, which is one of those with a grate front so I can see the fire burn. It is almost as good as a fireplace. The logs are unhewed outside because I like the rough finish, but inside the walls are perfectly square and smooth. The cracks in the walls are snugly filled with "daubing" and then the walls are covered with heavy gray building-paper, which makes the room very warm, and I really like the appearance. (Stewart 1914, 63)

The Woody family settled into their new life. No doubt they met and learned "all about" their neighbors, and those neighbors probably returned the favor. Roundups, brandings, and other events took on the trappings of social occasions. Wyoming had been holding rodeos since 1872, and it is likely that the Woodys attended or participated in what became the Cowboy State's official sport. Rodeos were a common event, and they brought together neighbors who shared information, gossiped, and generally put aside their collective sense of isolation.

The children had to be schooled. Uinta County's first school, which dates back to 1874, was established for the children of Ute Indians. Because young people were important in helping to efficiently run the ranches, time spent in the classroom often took second place to taking care of the ranch. Branding, planting, and harvest times all likely emptied the classrooms.

In 1900 (thought to be the first growing season in Wyoming for the recently acquired Woody family ranch), A. B. raised a bumper crop of wheat. When the grain was entered into competition at the Wyoming State Fair in Cheyenne,

Woody family history recalled that Azariah's wheat took first place. This auspicious start to Azariah's dry farming of wheat was short-lived. Subsequent to the bountiful wheat crop of 1900, good crops occurred roughly three years apart. This exemplified the difficulty of dry farming in an arid environment at such high altitudes.

The Woody family relied heavily upon their ability to grow most of their own food. Despite the harshness of the climate and its unpredictability, the family was able to raise sufficient amounts of both livestock and garden-reaped produce to survive. Hunting of the region's abundant wildlife provided an important supplement to the family's diet; it was not a sport, but a matter of survival.

The Homestead Act made it possible for less-affluent people to have a shot at achieving the "American dream" of a farm or a ranch of their own. But turning unimproved land into a working homestead was not always an easy process. Picking the right place was important. In a letter from April 18, 1909, Elinore Pruitt Rupert (she had not married Mr. Stewart by the time this passage was penned) wrote about picking her homestead site: "I have not filed on my land yet because the snow is fifteen feet deep on it, and I think I would rather see what I am getting, so will wait until summer. They have just three seasons here, winter and July and August."

On November 21, 1919, Azariah received a patent for 320 acres of land under the 1862 Homestead Act. The parcel became known as the Woody Bench. The Homestead Act required that the applicant maintain five years of residency on the subject parcel, as well as make improvements to the land over that period of time. This meant that in 1914, Azariah began his application process for the government-granted parcel of land he intended to ranch alongside the two hundred acres purchased from Joe Hatch. As part of his "proving up" the homestead application, infrastructure—such as another home, fences, fields, forest clearings, wells, spring developments, or similar improvements—was required.

The Homestead Act was modified in 1909 to increase grants of land from 160 to 320 acres, as the smaller acreage was insufficient for farming in the West, absent irrigation systems. Like most homesteads of the period, patents were issued to ranchers who already owned land but sought to expand the size of their holdings.

Prime homesteads had reliable water and bottomlands. Water was the key, as it sustained people, livestock, and crops. A stable water source allowed the homesteader to have an advantage in using the open range surrounding the homestead grant. Water rights were every bit as important as the land itself. A pasture with abundant grass was of little use if the water supply was insufficient to keep the livestock there long enough to utilize the available fodder. An 1894 Government Land Office map depicts that the parcel chosen by Azariah met the condition of having a substantial watercourse running through it. It was a good place for a homestead.

In 1911, Uinta County was divided into two counties: modern Lincoln and Uinta. The county seat was the town of Evanston, a Union Pacific Railroad town that sported a roundhouse. Evanston was somewhere over forty miles away from the Woody homestead, and that distance included many miles of rugged mountainous terrain. It was not an easy or fast trip by either wagon or horse. It was natural for the communities of Fort Bridger, Lyman, Mountain View, Lonetree, Carter, and the far-flung ranches to look more to each other than to the county seat located on the other side of a mountain near the border with Utah. The isolation of the Bridger Valley worked well to bring its inhabitants together.

The Woodys could have continued to focus on ranching, and clearly they were well on their way to becoming one of the more successful homesteads in the community. However, there was something in John D. that would not let the status quo be sufficient. He wanted to do something different. John D. Woody had a dream. It is to that dream the next connection takes us.

Chapter 2

The First Generation: John D. Woody's Dream

You know, if we keep trying we'll probably end up serving all of Wyoming.

—Attributed to John D. Woody, speaking to Howard Woody

It is not clear exactly when John D. Woody's dream came to him, or what may have actually set him upon the course to build a telephone company. There are suggestions that John D. may have begun working on "farmer lines" as early as 1900, when he first connected two ranches together. In 1910, he helped connect another two ranches. The newspapers of the time contained articles about the rapidly developing long-distance systems. Telecommunication generated significant excitement, and it had clearly passed the stage of being a curiosity or a fad. It was an up-and-coming utility. At some point, John D. Woody's experiences connecting his neighbors, coupled with the national enthusiasm for the emerging telecommunications industry, metamorphosed into a dream that became his life's calling.

The dream was simple in concept: John D. intended to connect his neighbors' ranches and the small communities of the Bridger Valley area, with an eye to including the ranches and communities in nearby Colorado and Utah as well. He wanted to build a modern telephone exchange deep in the heart of America's Outback. In execution, his vision was a complex and ambitious achievement. For John D. and generations of the Woody family yet to be born, the dream established a family mission and traditions of customer service that have survived into a new millennium.

In 1914, at the age of thirty-one, John D. actualized his dream and began building upon his vision of linking together neighbors in the greater Bridger

Valley area. Telephone companies had been springing up like weeds in rural America, and entrepreneurs would cobble together independent telephone services in these areas. Small communities created cooperative exchanges that linked together both people and businesses. Often, the exchanges revolved around connections between businesses within a town or community, such as linking a doctor's office with the local druggist. In more rural areas, ranchers linked up with their nearby neighbors on lines that often were hung from the fences in the fields. All of these systems were local, and they were generally unable to connect into the burgeoning system of long-distance lines.

AT&T and the major city exchanges mostly ignored rural Americans. AT&T was the largest phone company at the time. Its former parent company, American Bell Telephone Company, sold itself (for business purposes) to AT&T in 1899. Telephone company executive W. H. Barker summed up the belief held by the larger telephone companies of the time period: rural customers were burdensome. He called farmers "troublesome customers" with exaggerated views of their rights. Barker recommended overcoming their tendency to "bumptiousness" by constant educational efforts. Through such efforts, he continued, "there should certainly be some way of checking the senseless gossip that is carried to interminable lengths on some rural lines" (Casson 1910, 286). For a while, AT&T encouraged farmers to extend their lines toward AT&T exchanges, because the telecommunications giant hoped to expand its network at little or no cost. Around 1910, the competition with the independents slowed, and so did AT&T's interest in acquiring rural systems.

AT&T held a virtual monopoly on long-distance services, and they were focused on linking population centers and providing communications for the profitable high-density markets. AT&T often waited for outlying communities to build their own infrastructure; they would later snap up these rural exchanges for incorporation into AT&T's regional systems.

Gobbling up the independents was easy for the large telephone company. AT&T, working alongside banking magnate J. P. Morgan, effectively leveraged a credit crisis for the independent systems that AT&T wanted to acquire. The credit woes that Morgan helped create allowed the telephone giant to buy up the distressed companies' stock and consolidate them into state and regional companies controlled by AT&T (Brooks 1976, 132–133).

AT&T further controlled the independents (and picked up many of the most profitable ones) by buying up equipment-supply corporations, as well as restricting the independent systems' access to AT&T's long-distance services. AT&T's purchase of the Kellogg Company (at the time a major supplier of telephone equipment that independent exchanges puchased), was used to sell AT&T's competitors equipment made with parts that violated Bell's patents.[1] When Bell won the subsequent sham court battles for "patent infringement," many independent exchanges were forced out of business, as they were unable to bear the expenses of fighting a major company like Bell during a long court case (Fischer 1992, 45–46).

This is the Kellog switchboard purchased by Union Telephone.

AT&T/Bell had become a goliath, intent on establishing a worldwide monopoly for telephone service. Some called AT&T the "Bell Octopus." Union Telephone Company was one of many independent exchanges that purchased its switchboard from Kellogg. Had Union been profitable during that early period, they too might have fallen prey to the bogus, patent-infringement scheme.

Farms located near the expanding system of transcontinental telephone lines often were not allowed to tap into the AT&T system. Despite their proximity to a long-distance line, individual farmers sometimes had to drive miles to a population center to place a long-distance call at a pharmacist, a doctor's office, a local phone office, or a pay phone. AT&T had no compunction whatsoever regarding the inconvenience for rural farmers; at that time, AT&T was not planning to fill in the huge gaps in phone service in rural America. With the major phone system

[1] *While the principle characters and corporations involved keep changing, versions of this scam have continued into the twenty-first century. Patent infringements are alleged in apparent attempts to receive a lucrative settlement to keep the matter out of court and spare the defendant the costs of an expensive and complex court battle.*

leaving long-distance-service vacuums throughout much of the nation, small systems sprang up by necessity to meet rural America's long-distance-service needs. The inhabitants of America's Outback were not going to let corporate decisions keep them forever trapped in a telecommunications dead zone.

John D. Woody—either by good fortune or a shrewd understanding of circumstances—picked an opportune time to start his independent phone company. The long-distance system was developing rapidly, and in 1913, in order to avoid antitrust action by the government, AT&T entered into the Kingsbury Commitment (Brooks 1976, 136–137; Fischer 1992, 46–47). In a letter laying the foundation of the commitment, Nathan C. Kingsbury, AT&T vice president, promised to:

1) Divest the company of Western Union Stock (which effectively ended AT&T's control of its former competitor).

2) Purchase no further independents without Interstate Commerce Commission approval.

3) Arrange for other companies to use the lines of the Bell System (which, through AT&T's regional outlets, provided long-distance access for the independent phone companies).

This apparent reversal of policy by the Bell System slowed its inexorable march toward a national monopoly, but it did not put the matter to rest. Independent exchanges, such as the fledgling Union Telephone, were guaranteed that the services they could provide to their subscribers would include long-distance services. Further, the giant Bell System could not snatch up (almost at will) these smaller companies through the devious tactics of the past. AT&T committed to submit such acquisitions to federal oversight. As will be seen, long-distance access to AT&T was not provided immediately.

The Kingsbury Commitment left John D. free to focus on putting his telephone exchange together without worrying that the giant over the horizon was eventually going to take over the clientele and infrastructure of Union Telephone Company. The commitment was an assurance that the biggest shark in the ocean was not going to gobble up the work of successful entrepreneurs. However, the big shark was still swimming in the same ocean with the little fish—and it had not lost its appetite.

UNION TELEPHONE'S FOUNDING FATHER

The hardworking John D. Woody enjoys a pipe in a rare, quiet moment.

Most of what we know about the personal life of John D. Woody comes from recollections by his family members. His son Howard Woody describes him thus: "Dad was the kindest, gentlest man I ever knew. You just had to admire the man that he was. He was just a kind person. He was just as sincere as he could be.... He heard about what the Bell System was doing, but didn't know anything about it. So he got kind of a correspondence course to find out more about it. He had learned to read and write and he had a tremendous memory."

Daughter Bonnielee Woody Alexander almost duplicates Howard's description of John D.: "[He was] the kindest, gentlest person you would meet in your whole life. I never, ever heard him raise his voice to one of us children. He was so kind that I could never stand to disappoint him." Bonnielee describes him as a calm-spirited person, one whom she loved to go out in the field with when he worked on the phone system.

Not many photographic images of John D. Woody exist, but in the few that do, one clearly sees the image of a hard-working man who was usually dressed in bib-coveralls or outdoor gear. John D. resembled Tom Joad, Steinbeck's fictional everyman character that he wrote about in *The Grapes of Wrath*. "His eyes were very dark brown and there was a hint of brown pigment in his eyeballs. His cheekbones were high and wide, and strong deep lines cut down his cheeks, in curves beside his mouth. His upper lip was long, and since his teeth protruded, the lips stretched to cover them, for this man kept his mouth closed. His hands were hard, with broad fingers and nails as thick and ridged as little clamshells. The space between thumb and forefinger and the hams of his hands were shiny with callus." In Woody family photographs we see John D. with his cap placed upon his dark hair and his jaw firmly set. John D. appears as a serious man with a ruddy complexion. His is the physique of a worker, with strong hands and bandy arms and legs.

THE BIRTH OF UNION TELEPHONE

On January 28, 1914, John D. Woody's dream became a reality when Union Telephone was formed from the Mountain View Mutual Telephone line, the Smith's Fork Mutual Telephone line, the Black Forks Telephone line, and the Lonetree-Linwood Telephone Company.

One of John D. Woody's earliest tasks involved connecting the crude "farmer lines" of two ranches together. Once this was accomplished, John D. arranged to connect additional lines. In short order, this simple connection grew to a party line of roughly a dozen telephones. Since there was no switchboard at this time, the party-line customers contacted each other by turning the crank on their ponderous magneto telephones. Each rotation of the magneto handle created a ring throughout the system. The customers each had a specific number of rings, which indicated to the party-line members exactly who the intended recipient of the call was. For those who appreciated peace and quiet, the negative side of this arrangement was that all of the phones in the system rang every time there was a call. For those who appreciated the chance to listen in on their neighbors, it was a good thing that they were always alerted when someone on the party line had a call.

John D. managed to link together at least six isolated exchanges and individual ranch lines. His son Howard recalled the idea behind the name of the company: "The people in Robertson had built a telephone line on their own just for the ranches around there so they could call each other. Then there was another one at Fort Bridger. Then there was another one at Linwood, then Manila, then Lonetree, Burnt Fork, and McKinnon. Bringing all these together was Union Telephone." The new telephone company was built in baby steps. John D. had started out by helping to build upon and maintain the existing ranch lines. Then he was slowly able to interconnect the ranch lines at his own expense. Lastly, he connected the ranch party lines to the Mountain View exchange.

It was not an easy task to get individuals to join the exchange, even at a time when rural dwellers were installing telephones and paying higher rates than city dwellers (Fischer 1992, 93). Rural telephone systems tended to be small cooperatives, where the subscribers shared both the costs and labor necessary to build and maintain the exchange. Equipment used in these rural systems was often dated and approaching obsolescence. Lines were generally open and unshielded, and they were subject to fallen branches, drifting snow, and ice shorting them out. Some lines used the existing barbed wire fencing that crossed the backcountry. Many such rural exchanges did not even have a rudimentary switchboard and continued to function as large party lines.

John D. soon rented an office building, where he intended to install a switchboard. Switchboards were expensive and, once ordered, took several months for the manufacturer to fabricate. Using his ingenuity, John D. first ordered a switchboard from AT&T's Kellogg subsidiary. Then, during the months he awaited the delivery, he constructed his own switchboard. He manufactured a four-line switchboard incorporating a telephone from B-R Electrical and Telephone Manufacturing Company of Kansas City. For connectors, he used thirty-forty-caliber cartridges for the plugs, and thirty-thirty-caliber cartridges for the jacks. A replica of this homemade wonder adorns a wall in the present-day Union Telephone board room. John D. did not realize it, but this innovative device represented the first of many clever inventions his telephone company would produce.

The presence of a switchboard meant that operators were necessary to connect the customers to one another. When the Kellogg board arrived, it had the capability of connecting fifty lines. Millie Davis was hired to run the switchboard. As the company slowly grew, John D. eventually built an office in Mountain View and placed the central switchboard there. All of the "earnings" of the company were sunk right back into the business. The new office was a workplace for the operators. A cot was placed near the operators' station so that operators working long shifts could grab a few winks of sleep within earshot of the switchboard.

The telephone company provided more than communication services to its customers. The operators served in the role of the town crier, making announcements on the telephone system that were of importance to the community. Later, this was expanded to include daily announcements about what movie was showing at the nearby cinema. Preceding any such announcement, the operator gave a special signal consisting of three long rings. This alerted the customers that a public announcement was imminent. The operator waited and listened for the customers to pick up before making the announcement.

This collage showcases the replica of Union's first switchboard. John D. Woody converted a B-R Electric Telephone into a switchboard that used brass cartridges for contacts.

There were challenges in keeping such open-line systems running in a temperate climate; the challenges were even greater in the legendarily harsh climate of the Rocky Mountain West. Wyoming's severe winters and summers tested the limits of the technology, and it is a wonder that the crude systems functioned as well as they did. A simple service call for a downed line might begin a several-day ordeal, requiring the repairman to travel into the teeth of a blizzard or navigate through a sea of mud to reach the stricken line.

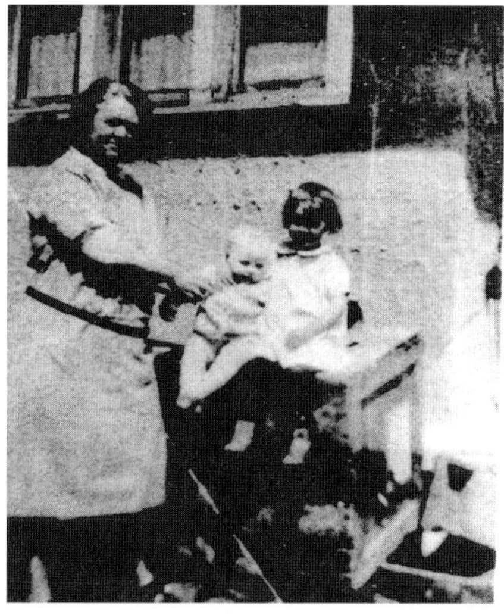

Working mom Millie A. Davis with her children, Wes and Phyllis, outside of Union's Central Office in Mountain View in 1928. *Photograph courtesy of Carol Davis*

Given the substantial costs of building rural systems (as well as rural subscribers'

The first Union Telephone Company office initially housed the Woody family. Later, operators would stay in the structure overnight during their long shifts.

resistance to paying enormously higher monthly rates for service), it is not surprising that the larger phone companies were hesitant to build telephone exchanges in the Wyoming frontier. While the service was vital, there were strong financial disincentives, and many of the rural companies withered on the vine.

In 1915, Governor Kendrick signed into law an act that established a public utilities commission in Wyoming (Larson 1990, 389). The act was among many passed that year in a broad legislative agenda that (among others) included workmen's compensation, women's rights, game and fish regulations, and farm loans. This meant that henceforth, utility companies such as Union Telephone Company would be regulated from Cheyenne, Wyoming. The Wyoming law established the first of many regulators the fledgling company had to satisfy.

John D. was not a member of the initial board of directors for the company. This may strike the reader as unusual, but this situation fit well with what we know about John D. Woody's personality. Initially, John D. was more interested in building the company and maintaining service than running the

Wyoming's harsh climate often disabled both telephone and power lines. This image is from 1917. *Photograph courtesy of the Wyoming State Archives*

business end. As time went on, it became clear that the decisions of the board (who did not seem to understand the workings of the systems in the field) sometimes affected John D. Woody's abilities to build and maintain the system.

John D. spent much of his time negotiating to obtain subscribers, often offering stock in the company as an incentive to join the Union Telephone exchange. This easy-going man must have learned to become an expert at the soft sell. For the most part, the initial subscribers were the owners of the company in a cooperative venture. Stock was used to pay for the corporate legal work and for contracted services, such as surveying and engineering.

Residents of Battle Mountain, Wyoming, pose atop a utility pole that is almost buried by snow. *Photograph courtesy of Grand Encampment Museum*

Howard describes his father's activities thus: "He never realized it was worth anything. He just was trying to help the people in Bridger Valley and make a living at the same time he formed the company. He sold shares of stock to start the company at fifty dollars a share and they spent twenty-five dollars for the telephone and spent the other twenty-five dollars trying to build a line to them. . . . He always ended up short and always had to dig into his own pocket and come up with whatever he needed to buy to build it. So he never really owned [the company] and it was that way right to the end."

Through perseverance and hard work, the young entrepreneur eventually managed to pull together an ambitious system of over three hundred miles of line. These lines were connected together through a central switchboard, originally placed in the Woody home in Mountain View.

Union Telephone's first board of directors consisted of ranchers John Briggs, George Stoll Jr., and Tom Welch. Including influential ranchers on the board was a double-edged sword. These ranchers—and others—helped John D. build the cooperative, but they knew very little about maintaining the system.

John D. found himself in an increasingly difficult situation as time progressed. It became harder and harder to get approvals for measures necessary to operate the company. The board did not understand how the telephone system worked, nor what was needed for John D. to keep it running. But John D. didn't hold sufficient stock to control the company. It was a problem that vexed him for years.

John D. himself had to do the majority of the backbreaking work involved in setting up poles and stringing new wire. The poles were cut on National Forest lands and then stripped of bark and treated prior to being set in the ground. John D. had originally used untreated poles; however, these quickly rotted and needed replacement. As a solution, John D. set up his own treatment facility where creosote, a tar created by burning coal and wood in a furnace, was used to preserve the poles. This preservation process added years of usability to each pole. It was common for the sections of poles buried in the ground to decay faster than the exposed portions. For the untreated poles of older telegraph lines that were converted to telephone lines, Howard tells of

This cartoon by D. Green in the *Casper Star-Tribune* pokes fun at a situation that actually happened in the Union Telephone offices. *Image courtesy of the Casper Star-Tribune*

how he and John D. dealt with them by pounding the poles deeper into the ground. "We'd just keep setting them another two or three feet in the ground until they were only about ten feet high."

The work of stringing the lines continued throughout even the worst weather, and Wyoming can throw some pretty hard weather at a fellow. When stringing wire to poles, John D. never used a ladder; instead, he scampered up and down the slippery poles using the lineman's spikes he had purchased and strapped to his legs. Digging the spikes into the pole with his feet while grasping it firmly in his strong hands allowed John D. to adroitly ascend and descend the wooden poles. John D. became extremely adept at performing his acrobatic work. The young man did much of the line work alone, occasionally receiving help from the ranches to which he provided service.

John D. helped his neighbors when he could. He sometimes served as a grocery delivery service, bringing needed supplies to customers while on his maintenance runs deep into the backcountry. On one occasion, the Union Telephone office resounded with the peeping of roughly a thousand chicks that had been left at the post office. The good neighbors in the telephone company office kept the chicks safe and warm until the owners retrieved them.

AN UNEXPECTED JOURNEY

When Azariah brought his family to Wyoming, his wife Isabell was in poor health. In 1915 or 1916, as her condition worsened, she sought help from a doctor who diagnosed her as having tuberculosis. Tuberculosis (short for *tubercle bacillus*) was commonly called the Great White Plague or Consumption in the early twentieth century, and it had a mortality rate between twenty-five and fifty percent. While the climate of Wyoming is suitably dry, the elevation was too high—and the air too thin—for someone in her condition. The doctor told the family that Isabell needed to recuperate by the ocean. The family chose to take Isabell to Oregon, which had opened its first private tuberculosis sanatorium in 1905. The sanatorium employed doctors who were familiar with the most modern treatments of the disease.

However, it was not the hospital or doctors that prompted the family to move Isabell to Oregon. Isabell's doctor had identified Oregon as a place that had a climate that might improve her condition. Isabell was not hospitalized during her stay in Oregon, but received home care from her family and local doctors.

Lottie Woody Simons in her garden in Sweet Home, Oregon, where she, John D. (her brother), and Burt Simons started another independent phone company to service the lumber industry.

While John D. and Lottie took Isabell for treatment, A. B. remained in Wyoming. At that time, Azariah was in the second or third year of proving up his application for a homestead patent. Had he abandoned the homestead, the family would have lost the land they had worked so hard to acquire, as well as all of the improvements they had made on the property. Yet the family needed funds for the move. The solution was to keep ranching the homestead-patent land and sell off the property they owned outright. Before John D. and company left, the family sold Joe Hatch's former ranch to fund the move to Oregon.

The need to relocate to the Pacific Northwest left John D. with another problem: What was to be done with the Union Telephone Company? John D. was not the kind of person to let his mother go through her ordeal alone, but he had also committed to completing the Union Telephone exchange. The Woody family philosophy has been to ensure that the customers come first. His solution to these conflicting responsibilities was to ask Earl Stoll, son of Union Telephone Company board member George Stoll Jr., to take over the work of maintaining and expanding the system while John D. cared for Isabell (Davis & Walker 2004). John D. was a man who knew his duty to both his family and the company. The chosen solution allowed him to stay faithful to both.

The Woodys packed their belongings into a wagon and began the long trip to Oregon. Once there, they established a new homestead on a wooded parcel of land near the community of Sweet Home. Their new home was roughly ninety miles from the ocean, nestled in the foothills of the Cascade Mountains, north of Eugene, Oregon. Lottie and John D cared for Isabell. After Lottie married local resident Burt Simons, he became Isabell's third caregiver.

While in Sweet Home, John D. decided to try his hand at building a new telephone company in Linn County. He had brought his homemade switchboard with him, and soon he was busily connecting neighbors again. He seized the opportunity to make his dream come true in his new surroundings. Once again, it took hard work, negotiating skills, and a little luck to build the exchange. Burt likely helped John D. with the labor. As a member of one of the pioneer families of Linn County, Burt's connections and influence were indispensible to establishing and building the new telephone company.

After the family had been in Oregon roughly two years, Isabell's new doctors suggested that she would benefit from yet another change of climate. The physicians recommended that she be moved to a sanatorium in the drier climate of Arizona. Since they did not recommend one of Oregon's state-of-the-art sanatoria, but rather one in the desert, Isabell's condition must have gotten even worse than before. John D. turned over his new homestead and the new company to Burt and Lottie. For decades, the couple continued to reside near Foster Reservoir (Elliot 1991). Burt and Lottie ran the new telephone company for several years before selling it to the Lebanon Telephone Company.

John D. placed Isabell on the Oregon Short Line Railroad with a destination of Evanston, Wyoming, where Isabell's daughter Lona lived with her husband. Having traveled from Oregon to Wyoming, Isabell was able to catch the Union Pacific Railroad at the end of the Oregon Short Line in Granger, Wyoming. Her family met her at the depot in Evanston. Before moving for treatment in Arizona, Isabell wanted to see both of her daughters. In addition to Lona in Evanston, daughter Matilda was nearby in Kemmerer, Wyoming, a reasonable traveling distance from Evanston.

Lona Woody in later years.

John D. put his mother on the train, certain that once Isabell arrived in Evanston she would be in the good hands of his sisters. He then bought a motorcycle and took a circuitous route back to Evanston. John D. planned to continue on with his mother to Arizona after she had spent some time visiting with Matilda and Lona. He took advantage of the freedom of traveling on his new motorcycle and explored places he might otherwise have never seen.

Back in Wyoming, Isabell's condition deteriorated rapidly after her arrival in Evanston. On October 17, 1918, she passed away shortly after getting off the train. She was fifty-eight years old.

THE YANKS ARE COMING!

The United States entered the First World War in April of 1917. The war affected Wyoming and other rural areas in the nation in a similar manner. America's young men responded in droves to the call to arms. The ranches in Wyoming sent a large proportion of their able-bodied men to support the war effort. The women stepped up and took over the duties of keeping the ranches functioning, though there remained some reticence within the local culture about women taking over the actual herding of the cattle. The need for skilled workers in war industries further depopulated the Wyoming outback, as men and women alike were needed by the manufacturing sector. At a time when the labor pool was severely reduced, the agricultural products of the state came into high demand. Food, wool, and horse production swung into high gear on the national level, but Wyoming was having trouble maintaining the herds and flocks it already had before the start of the war. The war created a deep labor deficit in the Cowboy State. Isolated corners of the nation, like the Bridger Valley, struggled to keep up production rates because of the reductions in available laborers and the shortage of resources created by the war.

With so many men spread far and wide as they fought in a global conflict, the communications systems in the Wyoming outback took on special significance. Radios crackled with news, telegraphs brought information from the front, and the telephone brought parting messages from boys before they boarded their ships bound for Europe. More than ever, the party lines helped relieve the feelings of isolation felt by the remaining ranch families. Union Telephone operators helped keep their customers informed of war news and other current events.

There was great hunger for information about the war. Independent companies (like Union Telephone Company) played a significant role in shepherding rural areas through the war years. Party lines often shared letters written by the doughboys (soldiers) and squids (sailors) from their distant posts. Although crude by today's standards, the radio and the telephone gave immediacy to the news from the war zones. "Over there" seemed to be much closer than the maps suggested.

For years, there had been elements in Washington, DC, that argued for government control of utilities, including the telephone system. The needs of a country at war gave more weight to the arguments for nationalization. President Wilson was convinced to use his war powers to take over the railroads and the telephone systems.

It is not clear how this government takeover played out in the isolated rural phone systems of Wyoming. Union Telephone was a fledgling company with no employees of military service age. Union Telephone Company and other small, independent exchanges were mostly left alone by the government takeover of the telephone industry. It was a lucky thing for the small independents, as mismanagement by the government helped slow the overall growth of telecommunications. Connecting the small rural exchanges into long-distance service lines was a low priority for the Postmaster General.

For the most part, the post office could not handle the enormity of their new responsibilities, and they did nothing with the independent exchanges. The US Postal Service focused on managing the large telephone companies, but they managed them poorly. High installation fees kept new customers away and thus overall construction and expansion suffered. The US Postal Service was competent at delivering the mail, but proved incapable of running an efficient national telephone system.

For AT&T, the government takeover had profound effects. The Bell system was assigned the crucial task of maintaining contact between Washington and the American Expeditionary Force in Europe. Thousands of AT&T employees were sent to the Western Front to build and maintain communications systems. For example, the 406th and 411th Telegraph Battalions were participants in the 1918 assault on Saint-Mihiel. The 406th contained Bell employees from Pennsylvania, and Pacific Bell employees formed the 411th. General "Blackjack" Pershing started a women's telephone operating unit that contained roughly a hundred women fluent in French and English. The US Signal Corps suffered hundreds of casualties while working the thousands of miles of telephone lines that were built to sustain the war effort in France and Belgium.

RENEWING THE DREAM

After Isabell's passing, John D. returned to Mountain View and began working for Union Telephone and the Woody ranch again. Azariah was in his mid-sixties and needed help running the ranch, as Earl Stoll had had about enough of running an unprofitable telephone company. Accordingly, John D. was able to return to his old job, doing everything to keep the slowly growing company up and running. During John D. Woody's absence, Earl built a telephone line from Lonetree to Mountain View, a project that was funded by Eugene Hickey, one of the most influential ranchers in Uinta County. The line followed part of the old Lodge Pole Trail, a Native American trail that was cut into the ground by the passage of travois traffic traveling between Roosevelt, Utah, and Kemmerer, Wyoming. The line was built so that Hickey could talk with the Taylor Ranch. The line route was not the best, and it was maintained—with some difficulty—by Hickey's sons.

John D. Woody and Mary Ellen Martin Zufelt Woody in front of their home in Mountain View, Wyoming. Mary Ellen worked as the primary telephone operator for the newly formed company in the late 1900s and for some years after.

The armistice that ended the war did not put an end to the suffering. Cessation of hostilities brought a welcome peace, but it also coincided with the influenza pandemic of 1918. Troops returning home from the battlefields helped spread this especially virulent strain of swine flu that killed millions worldwide. Within the United States in 1918, roughly 675,000 deaths were attributed to the disease. The American Outback was not remote enough to be left untouched by this tragedy. Along with many others, Betty Hickey, wife of Eugene Hickey, was a casualty of the pandemic.

John D. settled back into his routine of ranching and building and maintaining the telephone system. However, there was more to his life than building his dream telephone company. John D. started a family of his own. On April 8, 1922, he married Mary Ellen Martin Zufelt (also known as Polly),[2] a twenty-one-year-old recent divorcée with two daughters: Goldie, age two, and Verla, age five. John D. and Mary Ellen met at a boarding house where Azariah was staying. Coincidentally, Azariah was in a relationship with Mary Ellen's mother, Sarah, whom the elder Woody married.

According to John D. Woody's daughter Bonnielee, Azariah did not approve of the match between Mary Ellen and his son. A. B. knew that Mary Ellen was very much the opposite of his soft-spoken, easy-going son. In June of that same year, Mary Ellen delivered to John D. his firstborn child, Howard D. Woody. From this point forward, the Union Telephone Company was on the path to becoming a family business. The children in Union Telephone's second generation were brought into the world as follows: Verla (1917), Goldie (1920), Howard (1922), Lottiebelle (1926), and Bonnielee (1928).

THE GHOSTS OF MOUNTAIN MEADOWS

John D. Woody's new mother-in-law was also his stepmother. Family lore maintains that shortly after Isabell's death, Azariah married Sarah E. Martin, a widow of lawman Samuel Martin. Samuel died a violent and sudden death near Burnt Fork, Wyoming, at the hands of an outlaw in 1912.[3]

Samuel Martin claimed to be one of the few survivors of the 1857 Mountain Meadows Massacre, one of the most infamous events in the story of westward

[2] *The Western States Marriage Index indicates that Mary Ellen married Samuel Zufelt in 1910. She would have been nine years old if this is accurate. The records of Brigham Young University place the date of their marriage at 1916, suggesting she married at age fifteen. Records often have errors in transcription due to illegible entries. Unless the records, taken together, document a 1910 betrothal followed by the actual marriage ceremony, the 1916 date seems the most accurate. It was accepted during those times to see young women married off in their early teens.*

[3] *The marriage records in the Western States Marriage Index indicate that Azariah and Sarah were married in 1912, the same year of Samuel's murder. However, the records show another marriage for Samuel Martin in that year to Emily Barr. This record likely is for Sam Martin Jr., who ran a boarding house. The records further show Emily Barr also marrying a James Martin in 1912. It is clear the records have errors. The family maintains that Azariah and Sarah had never met prior to Isabell's death, and that they were married after Isabell's death in late 1918.*

migration. The massacre occurred when a unit of the Utah State Militia attacked and besieged the Baker-Fancher wagon train from Arkansas and Missouri during the Utah War (Bigler & Bagley 2011). The emigrants (encamped near Cedar City, Utah, at a place called the Mountain Meadows) mistakenly believed Native Americans were attacking them. The Utah Militia leader, John D. Lee, attempted to convince the besieged emigrants that their situation was hopeless. Lee told the trapped emigrants that if they surrendered and agreed to stand trial on unspecified charges in Utah, they would be escorted to safety past the Native Americans. What the emigrants did not know was that almost all of the Native Americans they thought they saw were actually Utah Militiamen dressed up as tribesmen.

Samuel Martin was believed to have survived the Mountain Meadows Massacre of 1857. As an adult, he became a lawman in Wyoming. *Photograph courtesy of the Verla Vaughn family*

On the day of the surrender, September 11, 1857, the emigrants were divided into parties based upon their gender and age. In one group, a militiaman accompanied each male member above a certain age. In another group, militiamen escorted the women and children. The emigrants had not walked far from the relative safety of their wagon circle when the order was given to halt. At a pre-arranged signal, every Utah militiaman shot and killed the man he escorted. Just as suddenly, a swarm of militiamen disguised as Native Americans descended on the helpless women and children, brutally murdering all of the adult women and all but a few of the youngest children. Several infants and children were slaughtered in the next few minutes of unbridled butchery. The recent discovery of the bones of several of these children (and many of the women) at the massacre site speaks volumes about the wanton cruelty of the monstrous militiamen.

There was no quarter given to any adults among the hapless emigrants, and none were allowed to escape with their lives. The orphaned, surviving children were taken in and cared for by local families until the United States Government arranged for their return some years later. Perhaps as many as 140 emigrants were murdered, making this was one of the largest losses of life in the history of the immigrant trails, second only perhaps to the Mormon hand-cart disasters of 1856 (Bagley 2002).

The events surrounding the massacre have long been a matter of controversy. Throughout the years since the massacre, unrelenting investigations by government officials met roadblocks that were constructed by a well-organized cover-up. Much of the historical dispute revolves around how much Brigham Young (first president of the Church of Jesus Christ of Latter-day Saints, commander-in-chief of the Utah Territorial Militia, acting territorial governor, territorial superintendent of Indian affairs, and church prophet) knew, and whether or not he ordered the slaughter of the Baker-Fancher wagon train. Brigham Young did not censure any of the participants, despite having the authority to punish the killers under authority from his various offices. Young's apologists cast doubt on his responsibility for the act, but have difficulty ignoring his failure to punish—in any meaningful manner—those who carried out the massacre.

After a series of trials related to the killings, John D. Lee was convicted of the crimes. In 1877, he was executed by a firing squad at the site of the massacre. The condemned John D. Lee wrote his memoirs in jail and pointed an accusing finger at Brigham Young (Lee 1877). J. D. Lee went to his end claiming his adopted father, Young, had sacrificed him in a cowardly manner. The controversies did not end with Lee's execution, but continue into the twenty-first century. Revisionist histories attempt to keep alive the cover stories about the massacre that were fabricated by the murderers.

According to Woody family history, Samuel Martin was a toddler on that bloody day, and he was found wandering alone some distance from the massacre site. The Mormon couple that found the orphaned infant hid the facts about his background from their church and raised Samuel as their own child. Not knowing his real name, Samuel took the last name of his adoptive parents, Martin.

Except for an investigation involving DNA analysis of Samuel's descendants and those of the Baker-Fancher Party, there is no way to confirm this story. No documents attesting to the tale have survived. However, this is not the kind of story that might produce a written record. The Martins likely would have done little to draw attention to young Samuel and the circumstances of his survival. Samuel indicated that the Martins kept his discovery near the massacre site a secret and raised him as their own child. Most literate Mormons of this period kept meticulous diaries. Many of these important historic documents for the year 1857 mysteriously disappeared in what appears to be a cover-up of the events during the Utah War. Despite a lack of corroborating documents, the power of Samuel's unusual story echoes to this day.

Samuel met a quick and brutal end at a saloon in Burnt Fork, Wyoming. Reputedly, he was murdered on the back steps of the saloon during his search for an outlaw named Al Scruggs. This was likely Alvin B. Scruggs, born in Virginia in 1868. In 1910, he was a rancher residing in Fort Bridger with his wife and three children. The 1900 census lists him as "Albern," and the 1910 census lists him as "Albin," but the 1880 census names the same individual as "Alvin." He was the son of the Chief of Police of Lynchburg, Virginia. Al lived in Ogden for a period and was a carpenter there.

One of Scruggs's accomplices was said to have bushwhacked Samuel as the unsuspecting lawman prepared to enter the saloon. Some people speculated that the killing was a setup to continue the cover-up of the Mountain Meadows Massacre, as Samuel apparently was quite open about relating the tale of his adoption. This seems unlikely, but the creation of such a theory in light of the orchestrated cover-up surrounding the massacre was to be expected. That was the way it was for this cover-up: once that trail was taken, all kinds of unrelated events appeared to take on conspiratorial trappings. The ghosts of Mountain Meadows have not rested quietly, and they inserted their presence into the story of Samuels's life and final moments.

A FAMILY BUSINESS

John D. Woody's wife and daughters helped to run the switchboard from their home. To make a call, a customer cranked the magneto on their phone, which sent a charge along the line that notified the operator. The operator picked up the call by plugging her telephone's jack into the appropriate plug that was connected to the caller. The operator then asked the caller for the name of the customer intended to receive the call. The operator worked a key with one hand,

and used the other to crank a handle on a generator to provide power to ring up a customer (the telephones of the period worked on dry-cell batteries). The operator then transmitted a coded series of rings (such as one long and two short, or one short with one long followed by one short, etc.) to the receiving phone. Each telephone user had to remember and recognize their ring code, or they were certain to incur the wrath of the members of their party line whose call was incorrectly intercepted. When the telephone's receiver was picked up, the operator connected the calling customer with their desired destination phone.

During thunderstorms, lightning sometimes hit the lines, which affected both the switchboard and the lines. Howard weighed in on this matter with this recollection: "One of my earliest childhood memories that I can remember was watching the lightning flash on the lightning protectors right behind the switchboard when lightening hit the line. The switchboard was installed on the wall between the sitting room and the back bedroom. Dad cut a hole in the wall and slid the switchboard into the niche." Behind the switchboard were lightning protectors that Howard delighted in. He noted: "Watching the arc flash travel through those cylinders when lightning would hit the line—I was fascinated by the light show created by lightning traveling along the blocks. It was quite a light show."

While most of the calls into the switchboard were mundane, the life of an operator had its moments. Customers often called the operators to learn the latest gossip, find out about road or weather conditions, and sometimes to try and find the whereabouts of missing children or spouses. The operators sometimes shared cooking recipes and cures for sick children and livestock. Operators checked on the ill and comforted the isolated. They dispatched help to the afflicted when it was needed. After all, the operators were as much members of the local community as the customers. The Hello Girls were good neighbors first and foremost.

John D. Woody's oldest daughter, Verla, ran the business operations of the company for two years, helping to free him for other work. Daughter Bonnielee was drafted to help with the switchboard, working twelve-hour shifts. Howard started his training at an early age, going out with John D. to service the lines. The telephone lines of that time did not follow the roads, but tended to follow the shortest path. As the poles requiring service were often out in fields with no access roads nearby, there was a lot of walking involved in checking and maintaining lines.

For years, John D. did much of the work of building and maintaining the system on horseback. Howard recalled that at about age four, he rode on such a mission behind his father on the back of a tall, gray mare. "I can remember it sure looked like a long ways to fall." It was not until young Howard was fifteen or sixteen that he was allowed to help with the actual repairs for the lines.

Mary Ellen (or "Mrs. Woody") and John D. posing in front of the Dodge truck they used to ride the lines. Mary Ellen would first drop off John D. at one end of a long stretch of phone line, and then drive around to pick him up at the other end of the line where it crossed the roadway.

Howard, who started walking the lines at eight years of age, described the task of checking the lines: "We'd take turns. By the time I was helping him a whole lot we had a truck and the lines didn't follow the roads. So, somebody would have to get out and walk from where you saw the line last through a field. Might be one mile. Might be six miles. . . . He'd [John D.] pick me up and take me to the next spot. If I had a pretty good rest, he'd let me walk again. And if I didn't get a little rest, when I got older, he would let me drive to the next spot and he would walk." As Howard got older, he also carried the tools for his father.

MS. INFORMATION

Operators get some of the strangest calls. The following is an example of an actual call to the Union switchboard.

Operator: "Directory assistance. How can I help you?"

Caller: "I need a number in Utah, but I don't know where it is. Can you give me the number of the Price City Offices?"

Howard described one of the hazards of the job, which he observed one time while watching his dad from the truck: "I remember one time he was on a pole down the Alfred Hickey place and it was raining and lightning . . . the lightning struck him and knocked him down the pole. I was just a kid five or six years old and couldn't figure out why he was taking a nap out there . . . I was too young to know that Dad had been struck by lightning. . . . It was getting cold and time to go home. He finally came to and unbuckled his spurs and come and got in the truck. I was asking him why he was taking a nap out there. He said, 'I really wasn't taking a nap.' And that is about the size of it."

Lightning was not the only hazard young Howard learned about while by his father's side. He recalled: "Once, he sent me across Tom Welch's field to walk a line. In the middle of that land, there was a marshy gully that I had to cross. The deer flies were awful that year. They bit me and bit me. I can remember running as fast as I could to get away from those deer flies. They can kill you, you know. Anyway, when I came running out of the woods with my back, face, and arms all bloody and flies still biting through my shirt, Dad felt so bad. I was crying and yelling. He took off my shirt and took the shirt and me down to the creek and washed us both. We then went back to the truck and I started walking the next line on the other side of the road. The water felt really good and cool. That shirt dried to my back after that."

Even as a child, Howard Woody helped his father check the telephone lines on horseback.

When Howard was old enough, John D. taught his son to shimmy up and down a pole using lineman's spurs, a 1.5-inch-long spike attached to each boot with straps. John D. modified a set of conventional linemen's spike by cutting them in half and welding them back together in a fashion so as to fit young Howard's legs. Howard was exceptionally proud of his father's gift, and the pair of spurs became a prized possession. He explained: "Dad took an adult pair of spurs and modified them for me. He removed about one and a half to two inches of the metal strap, and then welded the pieces together resulting in a climbing spur that would now buckle just below my knee. I wasn't very old and my legs were too short to accommodate the adult spur that would have reached clear up my thigh. I learned to climb pretty young, and could climb right up to the top of the pole." By the age of twelve, young Howard was learning how to fix and maintain the lines just like his father.

"MRS. WOODY"

John D. and Mary Ellen were very different people—one might say opposites. John was calm and reasoned, while Mary Ellen was sharp-tongued and emotional. Daughter Bonnielee described her mother as "always angry" with a "fiery temper." Mary Ellen was raised in a large family, and supposedly her seven brothers spoiled her. By the time she married John D. Woody, "she expected everyone to do things just the way she wanted." Mary Ellen was hard on her children and constantly berated them. Howard recalled her as "the meanest woman on earth." The children liked to be around their father, but tried to avoid giving their mother opportunities to turn her sharp tongue on them by steering clear of her whenever possible.

Mary Ellen was not happy in the role of housewife. She hated to cook and inflicted culinary disasters on her family, such as pancakes burned on the outside and still-liquid on the inside. She more than made up for her lack of cooking skills by being a superior seamstress. Mary Ellen's sewing showed true art and tremendous skill, and her daughter Bonnielee recalled having the best clothes in school, all because of her mother's sewing prowess.

It was a great day for the family when Bonnielee began to learn how to cook in her home economics class. She quickly learned how to provide the family with more palatable meals than her mother was able to serve.

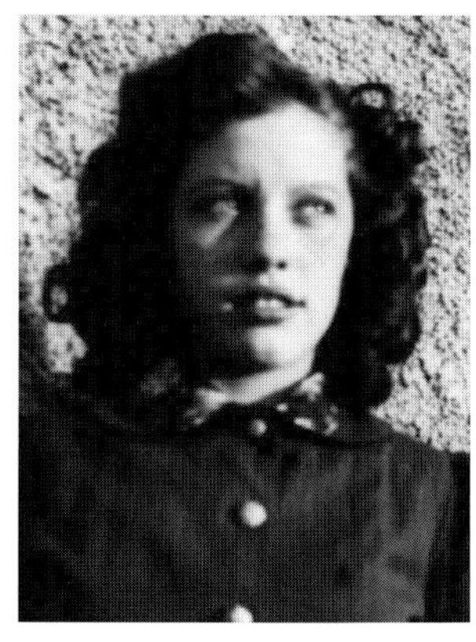

Bonnielee Woody

Mary Ellen was less confrontational with customers in her role as an operator. She also evidenced great compassion while serving as a midwife for the local community. Mary Ellen is reputed to have helped deliver many—if not most—of the babies in the

Bridger Valley over a period of several decades. She seemed to have shone in her role as a midwife and nurse, and she served as a competent operator. Her daughter Bonnielee speculated that Mary Ellen was not happy being married and a mother. Her level of dissatisfaction affected the entire family. Given her personality, Mary Ellen did the best she could in her roles of wife and mother. Bonnielee suggested, "She couldn't help but be what she was."

It is difficult to imagine the mild-mannered John D. interacting with the tempestuous Mary Ellen, whom he addressed rather formally as "Mrs. Woody." The couple stayed married for forty years, their union ending with John D. Woody's passing away in 1962.

GONE TO TEXAS

Sarah Martin Woody, the second wife of Azariah, died shortly after the two were married. *Photograph courtesy of the Verla Vaughn family*

Soon after Azariah married his second wife, Sarah Martin, she passed away. Azariah had lost two wives within a few short months. It must have been difficult to remain in a place with so many memories. Sarah was buried in the Burnt Fork Cemetery next to Samuel Martin. Sarah and Samuel reposed together in that quiet spot until just prior to the Second World War, when the Civilian Conservation Corps moved the cemetery to Lonetree. Azariah decided to return to the more halcyon climate of Texas to join his remaining family. He went back to Snyder and the staked plains after selling the homestead at Woody Bench.

Years passed with Azariah remaining in Texas, probably staying with family near Snyder. Eventually, he decided to head west and see his children again. Azariah first traveled to California in 1937 and visited Lona; he then traveled on to Sweet Home, Oregon, to visit Lottie. Azariah passed away while visiting in Sweet Home, and that is where he is buried. He never made it back to see John D. in Mountain View.

PARTY LINES: WHO'S LISTENING?

The company continued to grow, but slowly. In the early 1920s, the family moved out of the telephone office when they bought a residence on 4th Street in Mountain View. The company hired a local woman, Irma Johnson, as the main operator. Mary Ellen continued to take on operator duties. Irma lived in the telephone office building until she married Oscar Dahlquist and moved to his ranch.

Operators served to connect the customers to one another, and sometimes they were unofficial referees in the use of the exchange. Even in a small phone system, the job required a high degree of professionalism. To the subscribers, the operators were the voice and public face of the company. Good relations with neighbors was not only a pleasantry, it was sound business.

Early telephone systems had several homes or buildings sharing the same lines. A "party line" socially linked individuals. For example, all the ranches in a valley or on a side of a mountain might be linked together with a single phone line. This arrangement created a social network between neighbors. Like any community, there are those who are more sociable than others. The etiquette of such a network was very fragile, and perceived violations of the rules, selfishness, or rudeness could bring neighbors into conflict.

Flossie Vyvey, who worked as a Bell System operator, recalled an almost universal experience. "If you heard it ring and wanted to know what was going on, if you were kind of nosey, and who wasn't in those days, you'd go over and quietly lift the receiver, put your hand on the speaker thing, and listen to the old ladies talk, or whomever was talking." It seems that just about everyone who had a party line to their home has stories about their neighbor listening in on private calls, tying up the line, breaking into a call with comments, and on occasion taking over a call from the original caller. Even between the best of neighbors, sharing a single phone line often tried the forbearance of the party line members.

Once, when John D. Woody's daughter Bonnielee was serving as an operator, she discerned that when she rang up one of the telephones on a party line, several of the other parties picked up on the call, trying to listen in. She clearly heard the distinct clicks of the other receivers being lifted while the destination phone continued to ring. As eavesdroppers listened in on the call, they reduced the power available for the destination telephone to hear the incoming message. If a sufficient number of nosey neighbors picked up on a call, the volume dropped to such a degree that no one could hear the conversation.

After hearing the distinct sounds of several closely spaced clicks on the line when eavesdroppers attempted to listen in on a call, Bonnielee decided to let the interlopers have it. She spoke right up and called them all rude. An irate Bonnielee told the snoopers to get off the line so the destination party could hear their call. She knew it was a breach of professional etiquette, but Bonnielee had reached her limitations of tolerance for the practice of listening in on someone else's call. In retrospect, she recognized that the eavesdropping was a symptom of Wyoming's isolation. Any call, even one intended for a neighbor, presented a way to break up the monotony of the day and to connect with the wider world.

Early on, rural users learned that one of the benefits of the telephone was that it reduced the feeling of being separated from the community. Survey information from the early twentieth century revealed there was more telephone time spent socializing within the rural phone systems than those in the cities and towns. When private lines replaced party lines, the telephone companies had many customers who protested the loss of their *impromptu* social networks. For many rural residents, the private lines represented a return to the emptiness of isolation rather than the gift of privacy.

Dick Perue, a longtime resident of Saratoga and a former newspaper editor, recalled a story of a party line that had twelve users going from the top of a mountain to the valley floor. The user at the top of the mountain noticed a herd

MS. INFORMATION

Operators get some of the strangest calls. The following is an example of an actual call to the Union switchboard.

Operator: "Information—how can I help you?"

Caller: "How long do I cook a turkey?"

of twelve elk heading toward the valley. He knew his neighbors needed to fill their larders, so (after cutting an elk out of the herd), he put out an alert on the line that there were eleven elk heading down to the valley bottom. Shortly thereafter, the next family down the mountain (after they too cut an elk out of the herd) put out an alert that there were now only ten elk heading down the mountain.

This pattern continued throughout the day, but not all the party line members mentioned the number of elk remaining. They just passed along the information that the elk were still heading down the mountain. The elk herd was whittled away by the party line members as the animals attempted to descend from the mountain. However, someone along the route failed to mention they took more than one elk. No elk reached the home of the last member of the party line. The impatient lowermost member of the party line finally called up the others and asked what had happened to the elk. One can only hope that the luckier neighbors all shared elk meat with the last family.

Howard recounted a variation of this same story, this time occurring on the Henry's Fork River. The number of elk is the same, and the customer at the end of the party line also winds up without his elk. This story may be a form of a "rural legend" that recounts how members of Wyoming party lines helped to keep each other informed of important information, and how they sometimes did not share the resources. Knowledge about where to find elk (and the several hundred pounds of meat that would benefit a hungry family) was of great importance.

TIE HACKS

In the mountains that are scattered across southern Wyoming and eastern Utah, there lived a remarkable group of people who are remembered in history as "tie hacks." These hardy workers were mostly of Norwegian and Scandinavian descent. The moniker of tie hack owes its origin to the specialized work performed by these men, who spent most of the year harvesting timber and shaping the fallen wood into railroad ties.

Even in the most brutal winters, these tough-skinned souls ventured forth from crude mountain cabins to harvest wood or shape the downed timber into the dimensions specified for railroad ties. Hand tools were used to shape the ties, and working a broad axe to hew a log into railroad ties took great strength, endurance, and skill.

There are many photographic images of tie hacks traveling over the deep snows on long, wooden skis. They pushed themselves through the deep snows with a single pole, similar to how a gondolier navigates the canals in Venice. The tie hacks built small communities, and some even had stores, community meeting places, and an occasional saloon. All of these structures were built high up in the mountains at altitudes that most of the wildlife abandoned during the deep cold of winter. Theirs was not an easy life.

When the spring runoff flooded the drainage systems with a torrent of water from melting snow, the railroad ties were tumbled into the swollen waters to begin their several-weeks-long journey downstream. The nimble woodsmen rode and prodded the raw ties with their long-handled "peavies" (or cant hooks) through the winding river systems to collection points, such as Green River, Wyoming, or Fort Fred Steel. It is difficult to imagine the spectacle of tens of thousands of ties moving together as a giant, serpentine herd over churning rapids and snags, their nimble herders dancing precariously atop.

Injuries and deaths were common in this dangerous profession, but the job was how many people made their living during the early decades of the twentieth century. Howard summed it up: "A lot of them got killed doing it." The "tie drives" all ended at a collection boom near the railroad, where the ties were loaded into rail cars and sent for creosote treatment at the plant in Laramie. Howard recalled with admiration a time when he witnessed one of the great tie drives. It must have been something to see.

Howard was fortunate to have visited one of the tie hack camps with his father, and he shared this memory: "On the Black's Fork River, high up in the Uinta Mountains above the Meeks Cabin area, there was a tie camp. Malcolm McQuege ran the camp. Workers, many of them Swedes and Norwegians, came to the area purposely for cutting trees and making ties to help build the railroad. These tie hackers would harvest trees and flat cut them on two sides to shape them into railroad ties. All of this was done by hand. There were no machines to make those ties. It was fascinating to watch them shape those ties."

John D. recognized the importance of expanding the industries that Union Telephone served. A tie hack community in the Uinta Mountains, near Meeks Cabin, became a subscriber to Union Telephone's expanding system. As Howard put it, "It was just something he felt he had to do."

Bringing telephone service to other camps, such as the colorfully named Suicide Park, followed the success near Meeks Cabin. Local legend suggested Suicide Park was named after an injured tie hack despaired of receiving medical attention for a serious wound, and chose to cut his own throat on his razor-sharp broadax rather than suffer any longer. Interpretive materials at the Suicide Park cemetery named three individuals who were thought to have died at their own hands: Ole Olson (died 1928), Jack Rose (died 1928), and Charles Mattsen (died 1930). In order to have accomplished their respective suicides, Rose apparently used a revolver and Mattsen slashed his own throat with a razor. Olson's cause of death is not revealed, but it might be the origin of the local legend of suicide by broadax. The telephone connection to Suicide Park helped reduce isolation and made the remote camp more endurable for its occupants. Even so, these three individuals became despondent about further life in the outback, and took a dark escape route.

Getting to these remote camps created new challenges for John D. Woody. Unlike the ranch systems, there were no fence lines for him to follow. The line was strung from tree to tree up to the camp. It was a long and laborious process in some of the most rugged terrain the region had to offer. Only a dreamer could have had such dedication to string so many miles of line, alone, through the primeval forest, and then to service the lines going through places even the sure-footed horses had difficulty traversing. It was the sheer audacity of getting telephone service into the most remote locations that became a model that stayed with Union Telephone Company for decades to come.

LONG DISTANCE

In the 1913 Kingsbury Commitment, AT&T pledged to let the independent exchanges connect to their long-distance lines as an attempt to fend off potential antitrust action on the part of the Federal Government. However, that pledge did not guarantee immediate access. Long-distance lines were in their infancy in 1913. During the First World War, there was little expansion of long-distance services under the nationalized telephone system. It was not until 1929 that Union Telephone customers were allowed to connect with the outside world. Union was permitted to connect with a long-distance line that AT&T operated along the Union Pacific Railroad corridor. That line provided sufficient capacity for up to fifteen Union Telephone Company customers to access long-distance service at the same time. The promise of 1913 was finally fulfilled.

Access to long-distance service opened up new opportunities for Union Telephone. The company's main office eventually housed two payphone booths, which entertained a lively business and became a communication center for the community, especially for cattle buyers. With the availability of long-distance service, the buyers seized upon the convenience to buy, sell, and trade their stock from the booths in the Mountain View office. To the great delight of the operators listening in, the cattle buyers occupied the long bench outside the telephone booths and conducted long-winded, extensive haggling, trying to get the best deals for their livestock. The telephone office became a hub of activity and a center of general commerce.

IF WE MAKE IT THROUGH DECEMBER

Despite John D. Woody's Herculean efforts to make Union Telephone profitable, the company just could not turn that corner. If it were not for the ranch and the food grown there, the family would have been in dire straits. Howard recalled: "As a youngster, we were probably the poorest people in town. We grew a garden and my Dad always raised some pigs and usually had twelve or thirteen chickens. I never knew starvation, but sometimes you would get away from the table a little hungry."

Following the drought of 1919, the bottom dropped out on the livestock market. This factor, combined with reduced federal government spending at the end of the First World War, helped grind Wyoming's fragile economy to a screeching halt. Dozens of banks failed and folded (Larson 1978, 411–446). The Union Pacific Railroad let a third of its workforce go. Immigration into the state waned, and there were fewer homestead applications than in the peak year of 1919. Wyoming endured a ten-year head start on the Great Depression. It was not the ideal economic atmosphere in which to grow a vibrant telephone company.

The only bright economic light for Wyoming during the 1920s was the oil industry, which showed real growth. The Teapot Dome scandal, named for a Wyoming landmark near Casper, soon embroiled the industry in one of the most infamous government scandals. Under the Mineral Leasing Act of 1920, the leasing of federally owned minerals was regulated. The Teapot Dome scandal was a classic play-for-pay situation, in which Mammoth Oil leased federal minerals in Wyoming, and Pan American Petroleum was issued a lease in California by Secretary of the Interior Albert Fall. The secretary leased these strategic naval oil reserves to Mammoth Oil without competition, and Fall received substantial monetary compensation from Mammoth. When a congressional investigation started snooping around the deal, Fall began creating a cover-up. However, the secretary was convicted of taking bribes and jury tampering, while executives from Pan American Oil and Mammoth Oil walked away with nothing more severe than a contempt fine.

While big oil was apparently doing reasonably well (even after paying their fines), ranchers in rural Wyoming were running out of viable options to get them through the tough economic times. Many turned to supplementing their income with outside jobs. Children were required to step up and do more to keep house and home together. Some turned to moonshine production, as America's brief experiment with social engineering, called Prohibition, turned liquor into gold. Tourism in Wyoming continued to grow during this period, but unless one lived near a tourism Mecca like Yellowstone National Park, there was little effect on the lives of those in the average rural community.

When the Great Depression rocked the world in 1929, Wyoming had already been riding that bronco for almost ten years. The Spartan lifestyle necessary to get along in America's Outback served the people of the state well in times of privation. Ranches provided most of the food for their inhabitants, and with the economy facing another contraction, dry farming became an even riskier venture than it was during good economic times.

The supplemental food provided by hunting and fishing took on increased importance to those hoping to get through the hard times. Many inhabitants of America's Outback recalled having more deer or elk in their diet than beef or mutton during these lean years. The Wall Street collapse seemed like a continuation of hard times, only with more folks along for the ride. A series of droughts in the mid-1930s worsened the suffering nationwide. Howard's experience of leaving the table a little hungry took on national proportions.

The effects of the droughts in Wyoming were severe, but they were nothing like the ones during the Dust Bowl in Kansas, Colorado, and New Mexico, in which the topsoil was swept away in "black blizzards." The fugitive dirt piled up in city streets along the East Coast in an ecological disaster of unmatched proportions. An estimated twelve million pounds of dust fell on Chicago, with cities like Washington, DC, Boston, and New York also being buried in dust as the storms proceeded across the nation. The winter of 1934–1935 saw red snows falling in New England as a result of the ecological disaster in Oklahoma.

The Dust Bowl forced a mass migration of farm families to the agricultural fields of the West. One of these refugees was a young girl named Esther Pauline Gladwill from Pottawatomi, Oklahoma. She and her family fled the dust bowl, and—like many thousands of others—they took the Lincoln Highway through Wyoming and headed to the agricultural fields of California. At a stop not too far from Granger, Esther caught her first glimpse of the Uinta Mountains. She was so struck by the majestic vista that she resolved to return to visit them someday. She was unaware that those mountains had a role to play in her destiny.

John D. returning from a successful hunt with meat for the table. Hunting was not recreation for John D. Woody—it was a matter of providing for a hungry family.

Initially, telephone service did not suffer during the Great Depression. Inevitably though, the downturn in the economy resulted in reductions in the number of subscribers. The telephone was more of a luxury during this time period, and household money shortages eventually reduced the number of customers. As the depression deepened and revenues shriveled, it was necessary for telephone companies to reduce staff. AT&T attempted to do this mainly through attrition, and they continued to convert their system from manual to dialing. The Bell system saw a lot of red ink, especially in 1933, but they weathered the storm, waiting for things to improve.

The Great Depression put a stranglehold on much of the world's economy. It is difficult to imagine a time when the homeless and dispossessed citizens came together by the thousands and established "shantytowns" full of shacks, many of them just boxes or crates. These communities were called "Hoovervilles," named after President Hoover. The Hoover administration's austerity measures and monetary policies, combined with poor regulation of the financial industries, were perceived to be a large factor in turning what should have been a downtick in the economy into a financial collapse. Stocks plummeted. Banks collapsed. For those in Wyoming, the record number of failed farms was more significant than the collapsing banks. The federal government's response was a series of "New Deal" programs aimed to stimulate growth and get America "out of the red," meaning back to profitability. Some of these programs remained controversial in Union's centennial year.

During the Great Depression, the federal government became an employer of last resort. Many of Wyoming's youth went into the military or the Civilian Conservation Corps (CCC). Roughly 275,000 citizens were rescued from the despair of poverty by programs under the New Deal like the CCC. The CCC provided a small income, health care, education, and job training. The Roosevelt administration had appropriated the concept for the CCC from Nazi Germany, where German youth were put to work in massive public service projects to the benefit of the overall economy.

HARD TIMES COME AGAIN NO MORE

The Woody family suffered a major loss in 1935 when Goldie Woody died. Roughly three years before her death, she had been playing with a friend when an accident befell the little girl. She was struck in the head with a stone wielded by her playmate, Ed. After the accident she became very sickly, and despite the assistance of doctors in Salt Lake City, she never recovered. Bonnielee remembers her sister Goldie as being a kind and gentle soul. While her parents tended to Goldie during her medical treatment in Salt Lake City, the children were sent to stay with neighbors. Howard stayed with his lifelong friends the Dahlquists during this troubled time.

WHEN THE LIGHTS GO ON AGAIN

The Second World War officially began in September of 1939 when Nazi Germany invaded Poland, though portions of the conflict began as early as 1935. The conflict rapidly spread across the globe, eventually involving roughly one hundred million combatants. As many as eighty-five million people were dead by its end in 1945. The United States officially entered the ongoing conflict in December of 1941, when the Empire of Japan attacked the US Naval base and Army Air Force bases around Pearl Harbor in Hawaii.

Traditional history holds that the Great Depression ended with the start of the Second World War. That may be true for industrial areas, where excess manpower was siphoned to war industries and the building up of American armed forces. In the industrial heartland of the nation, the massive influx of cash into war industries stimulated the economy and put millions to work.

But rural Wyoming did not have excess people to offer, and the demands of the war affected many of Wyoming's inhabitants more than the economic collapse during the Great Depression. In Wyoming, the war siphoned off the young workers who were vital to keeping the ranching industry running. Both young men and young women were desperately needed for a total war effort, and America's Outback shortly became a haunt of the old and the very young. The vitality had been taken out of the region. It was as if rural Wyoming was holding its breath, waiting for the change that would restore the balance that once existed. If the Great Depression was a period of collapse that was followed by slow growth, the war years were a period of little or no growth and, in some places, shrinkage. Communities with military bases, such as Casper and Cheyenne, seemed to fare the best. With the hurdles of the Great Depression and several smaller financial panics and recessions to jump, it is little surprise that so few companies have survived to celebrate their centennial year.

The Union Telephone Office (right) in Mountain View, Wyoming, was a hub of commerce for the community. This photograph is circa the 1940s.

Like so many Americans, the Woody family was swept up in great events. In 1940, Howard chose to enlist in the National Guard. The United States was militarizing because of President Roosevelt's convictions that the nation inevitably would become entangled within the world war that raged in Europe and Asia. Lacking a large standing army, it was necessary to build up the National Guard to provide a body of trained men for the time when America was drawn into the conflict. For most of the men who were mustered in during the pre-war period, the goals of expanding the National Guard were nebulous. The unemployed youths needed work, and this was a job with room and board thrown in. The ranks of National Guard units across the nation swelled. In the aftermath of the attack on Pearl Harbor, Howard's term of service expanded from one year to the duration of the war plus an unspecified period.

Even with Howard pulling in steady pay, finances for the family were tight. To help make ends meet, Bonnielee took a job in another town. John D. had to sell all of the cows, and the ranch became a garden and chicken farm. It was time to tighten the belts another notch. The hard times that started with the drought of 1919 continued unabated.

John D. focused on keeping the company going through what must have seemed like an endless round of difficult times. The war made communication with the outside world seem more important. Customers gossiped about broader issues than cattle prices, rainfall, a new baby, and/or who had a little bit too much to drink the weekend before. During the war, Americans were hearing about places like Pearl Harbor, Singapore, Bataan, Midway Island, Guadalcanal, Tarawa, Normandy, Falaise, Stalingrad, and Dresden. These faraway places became part of the national conversation. There were the all-too-infrequent long-distance calls from servicemen and servicewomen. Sometimes a recording made in a port of call or at a base survived to bring home the voices of the servicemen and servicewomen. The static-laden voice of a loved one coming into the home from afar was a true gift to the families waiting for their return.

When the War Department wished to send news (which was almost always bad tidings) to the families of service members in the Bridger Valley, they contacted Union Telephone Company. John D. did not have to, but in service to his neighbors and his country, he took upon himself the burden of being the bearer of these official messages. It was a heavy encumbrance for the tenderhearted John D. He told his son Howard, "I always went and talked to them and tried to tell them everything the people had told me." Being the deliveryman for such horrific news was the part of running a telephone company that John D. hated the most. The entire time, John D. Woody (as a father of a soldier) hoped and prayed that he would not receive a message about the son of his who was fighting his way through the battlefields of Europe.

Our next connection takes us to the dark days of World War Two and a fortuitous encounter with destiny.

Chapter 3

The Second Generation: Making the Dream Come True

It is the most important thing I ever accomplished. I've done a lot of things in my ninety years. I always was doing things so I could just get back and build this company.

—Howard D. Woody

The second generation took John D. Woody's dream beyond the initial concept of creating a network of a few isolated neighbors when they began a region-wide expansion. It was during generation two's watch that the Union Telephone Company began to become real competition for other local exchanges, as well as the majors. The generation consisted of Howard D. Woody, his four sisters (Verla, Goldie, Lottiebelle, and Bonnielee), and his wife, Esther. As was typical for the times, the family business passed from father to the eldest son. But in this case the fourth Woody woman, Howard's own true love, Esther, became one of the important guiding lights for the company. Eventually, Howard's sisters moved on to lives outside of the company, leaving the bulk of generation two's story to be about Howard, his bride, and their children. The third-generation story tells mainly of their children, and not of the offspring of Howard's sisters.

Lottiebelle Woody (Howard's sister) worked as a telephone operator until she married.

Somewhat out of uniform, this photograph shows Howard when he was a buck private in the Wyoming National Guard.

Howard D. Woody was perhaps the most publicly visible member of the family. Images of Howard from his childhood show a smiling child. His signature Mona Lisa smile has served Howard well throughout his life. In early pictures of Howard wearing his military uniform with a cowboy hat, he looks like central casting sent him to play a role in a 1940s singing-cowboy movie. The same warm smile and bright eyes stayed with Howard, and they are still viewable on Union Telephone billboards, commercials, and their website. He looked exactly like the friendly guy he was, and when conversing with him, one was not disappointed in that regard. He had a disarming wit and an ability to spin just about any event into an interesting yarn. All in all, he lived a fascinating life.

A LOVE STORY

I never thought I'd live this long. I never thought I'd live past my wife, because she was my inspiration. She was the greatest partner a man ever had.
—Howard Woody speaking in 2013

A more in-uniform Howard during his training.

In 1940, Howard signed up for a year's enlistment, but the events at Pearl Harbor extended all such enlistments to the duration of hostilities, and then some. That extension of his hitch did not sit well with Howard. He did not appreciate that the terms of the deal between himself and President Roosevelt were changed without consultation. Even as far removed in time from that event as Union's centennial year, Howard still took the breaking of this agreement somewhat personally. However, Howard honored his part of the deal and served with distinction. His Wyoming National Guard unit trained hard and waited for an assignment to one of the theaters of action.

One day in 1943, Howard learned that the Navy was desperate to find men who knew how to splice cable.

They needed to warehouse huge stockpiles of naval ammunition far enough from the coast to avoid bombing. The Navy chose the forbidding deserts of Nevada, a place free of ships and a great distance from enemy attacks or sabotage.

Additionally, the high desert is a perfect warehousing climate. Each ammunition-storage facility encompassed hundreds (sometimes thousands) of buried huts or igloos. Munitions were stored underground in cement bunkers that were covered over with earth. The Navy needed people to hook each complex of igloos to central facilities where the temperature and humidity of the bunkers could be monitored.

Howard informed his superiors that he had been splicing cable since the tender age of six. He didn't realize it at that time, but events were set in motion that changed his life forever. Within a week of learning of Howard's ability to splice cables, his sergeant offered the man from Mountain View a new assignment. At first Howard hesitated, not knowing that the new duty station was at a naval facility in the desert. As Howard put it, "That ocean there never did attract me, but I told them it didn't matter what they do." Howard took the assignment and was relieved to find his new duty station was hundreds of miles inland in terrain that was familiar to him.

Howard was granted a two-week furlough to visit home, after which he was sent to Hawthorne, Nevada. This small town, roughly seventy-six miles southeast of Carson City, Nevada, is located in a remote basin almost completely surrounded by mountains. In Union's centennial year, Hawthorne remained in use as a storage facility for US military weapons and equipment.

Howard quickly realized that repairing the six-pair cables that required splicing was a simple duty. He set to work with a will. The normal workday was ten hours long. There were hundreds of storage buildings to serve, and even a hard worker like Howard could not splice cable fast enough to serve the needs of the expanding facility. Eventually, the Army had Howard train additional soldiers to help him splice cable.

Esther Gladwill Woody was the love of Howard's life.

It was on the job that Howard met an electrician named Gordon Gladwill. The two worked closely together and became friends. One night after work, Howard spotted Gordon at a local bar and they had a beer together. Gordon asked Howard to come home with him and have a "home-cooked meal." It was a tempting offer after so many weeks of military cuisine. Howard tried to back out of it for fear that Gordon's wife did not have sufficient warning of an unexpected guest for dinner. Gordon did not let Howard off the hook. Howard recounted the moment: "So I went home with him and I met his daughter. I thought she was the most beautiful girl I'd ever seen." For Howard, this chance meeting with Esther Gladwill changed his life forever. "I guess I realized the first time I saw her that she was the one," he continued.

Esther was born in Shawnee, Oklahoma, in 1924 to Gordon and Dorothea Gladwill. A photograph of Esther as a young woman shows a slim, petite beauty with a captivating smile. In this photograph, her thick, dark hair is adorned with floral arrangements. Her eyes betray a joy of living and an intensity that melted Howard's heart.

Early on in their acquaintance, Esther recounted to Howard the trip when her family was leaving the Oklahoma Dust Bowl. They traveled along the Lincoln Highway through Wyoming to the fertile fields of California. She told Howard that on the exodus from the Dust Bowl trip, she and her family were near Little America when they stopped to camp for the night.[1] To the south were the grandest snow-capped mountains. These majestic peaks of the nation's only east-west trending mountain range impressed young Esther so much that it became a dearly held wish that someday she would return to Wyoming to see

[1] *Little America started out as a motel that was built near the place where its founder was stranded during a blizzard. According to the company, the name the new motel appropriated was the same as the United States scientific outpost on Antarctica, which, by implication, must have had conditions resembling winter in Wyoming. In Union's centennial year, Little America is located on Interstate 80, several miles south of Granger, Wyoming. During the Great Depression, Little America was located on the Lincoln Highway near the town of Granger. The original motel site was destroyed in a mysterious fire and rebuilt along the interstate.*

the mountains better. Howard recognized from Esther's description that she was describing the Uinta Mountains, the mountains to the south of the Woody home, and the mountains for which Mountain View, Wyoming, received its name. It must have seemed like the two were fated to be together.

Esther and Howard Woody.

Esther and Howard hit it off right away. After first meeting Esther, Howard's loan to the Navy felt like pretty good duty. Howard and Esther made the best of his off-duty hours, as their whirlwind courtship led them in short order to fall in love and celebrate a hurried, wartime marriage. There were tens of thousands of wartime romances, but for Esther and Howard, they each had found the love of their life, and their marriage flourished for over sixty years.

About four months into their courtship, the day arrived when Howard received word that his detail to the Navy was about to end, and he had to return to his unit in the Army. Howard looked at Esther and said, "Are you going with me or are you staying here?" Esther replied, "I'm going with you." It was not the most romantic of proposals, but for Esther and Howard, straight talk with no nonsense was perfect.

During the next week, in September of 1943, the two traveled to Carson City, Nevada, and tied the knot. It was one of the happiest days of Howard's life. The Ormsby County clerk apparently thought they were too young and tried to dissuade the lovestruck pair from getting married. Howard remembers her saying: "You kids get married and then you don't stay together. If you're still together when this war's over, I want you to come back and see me." Howard replied, "Well, we're going to get married and when the war is over, if I'm still alive, we'll come back to see you." Roughly five years after this event, Howard and Esther returned to Carson City. It took some time to locate the clerk who issued their marriage license, as the woman no longer held that job. Esther and

Howard tracked down and visited the woman, proving that they were not a typical wartime pairing. Howard could not help but chuckle as he recounted this story. "She was delighted that we came back to see her."

The newlyweds made the most of their brief time together before Howard went to Oregon for training. The family set up a living situation for Esther in Mountain View. Howard did not have a place of his own, which required Esther to share housing for a while. When the couple was first married, they lived in the old homestead in front of John D. Woody's house. John D. and Mary moved to the telephone office to give Esther a house. Soon thereafter, Esther was expecting her first child.

Esther followed Howard to Oregon where he was in training, and she stayed with Lottie for a brief time until the couple could get an apartment. She waited tables for a while before returning to Mountain View when Howard's unit left for Europe. There she awaited Howard's return. When the weather was cooperative, Esther was able to look at the mountains that had so captivated her during her trip from Oklahoma on the Lincoln Highway. Shortly before Howard's unit shipped out to join the gathering invasion of Europe, Esther delivered her first son.

It was a wonderful surprise when Howard was granted leave to visit Esther and their new baby. Howard had to promise his commanding officer that he would return before the unit shipped out. Howard did not have to be told twice, and he immediately took the next train out and soon was back in Mountain View. When his leave was about to end, Esther escorted Howard to the train station. A gentle rain fell on their parting moments at the station, similar to a page from a script for a Hollywood movie of that era. But as Howard said: "It didn't feel like a Hollywood movie to me. She had become my life. We had sixty-one years together."

As Howard traveled to his uncertain future, Esther rolled up her sleeves and put on a brave face. The hard work of each day was punctuated by the latest news from the war front, as anxious hearts all over the nation listened for encouraging news on the radio, watched shocking scenes of carnage in newsreels, and prayed for their loved ones to return.

BLESS 'EM ALL: DARK DAYS OF WAR

I've seen so much wickedness in my life. The war was terrible. I've been fifty years trying to forget it.
—Howard Woody

The Second World War was much different for the telephone companies of the United States than the First World War had been. While the strategic importance of communication was understood, this time the government did not repeat the mistake of taking over the telephone companies and putting them under the Postmaster General. Resources were channeled first toward the war effort, but the war years also were a time of great need for long distance service. At first, long-distance usage quintupled; it then hit astronomical levels of demand as the conflict dragged on.

During the war years, communications-research laboratories came up with technological wonders, like improved radar and devices to determine both the origin points of the enemy's artillery shells and the rockets that were fired at US troops. While demand for services soared, overall profits for telephone companies remained flat. However, much of the groundwork for a postwar communications boom was laid during the war. The utility of a vibrant long-distance network was burned into the psyche of American consumers.

Howard began his military career in the 41st Infantry Division (the Sunsetter Division) as part of the Wyoming National Guard. Howard remembered many of his friends from the unit, including Harold C. Megeath, Albert E. Wall, Earl J. Brown, Leland D. Brown, Earl D. Bullock, Myrl D. Bullock, Sid Davis, James A. Felix, Oscar R. Fillin, Norman D. Johnson, Warren W. Meadows, Charles A. Meeks, Jewel F. Meeks, Frank S. Montgomery, Glenn F. Murray, Earl P. Robertson, Milton E. Rollins, Ellred J. Slagowski, Gerald Slagowski, and Owen J. Stewart. There were thirty-six recruits who signed up in Green River the same day as Howard.

The Sunsetter Division shipped out early in the war for a long and bloody series of campaigns in the Pacific theater. Battlefields like Buna-Gona and Guadalcanal waited for the 41st. Their experiences in some of the most foreboding jungles in the Pacific theater of war earned them the nickname "The

Jungleers." However, the same destiny that gave Howard his stint splicing cables for the Navy also kept him at home and allowed him to meet the love of his life. Prior to the 41st deploying to the Pacific, Howard was assigned to another unit and transferred as a non-commissioned officer into the 104th Infantry Division (the Timberwolf Division).

The 104th was a reserve asset division that trained at Camp Adair in Oregon. Their specialty was night fighting, and the night-vision goggles and thermal-imaging systems of today were merely science fiction during the Second World War. The ability to move and fight at night was a forte reserved to a few specially trained units. In 1946, Leo Arthur Hoegh and Howard J. Doyle wrote a history of the Timberwolf Division's activities during the Second World War, and it is an excellent account of the exploits of this remarkable Army division.

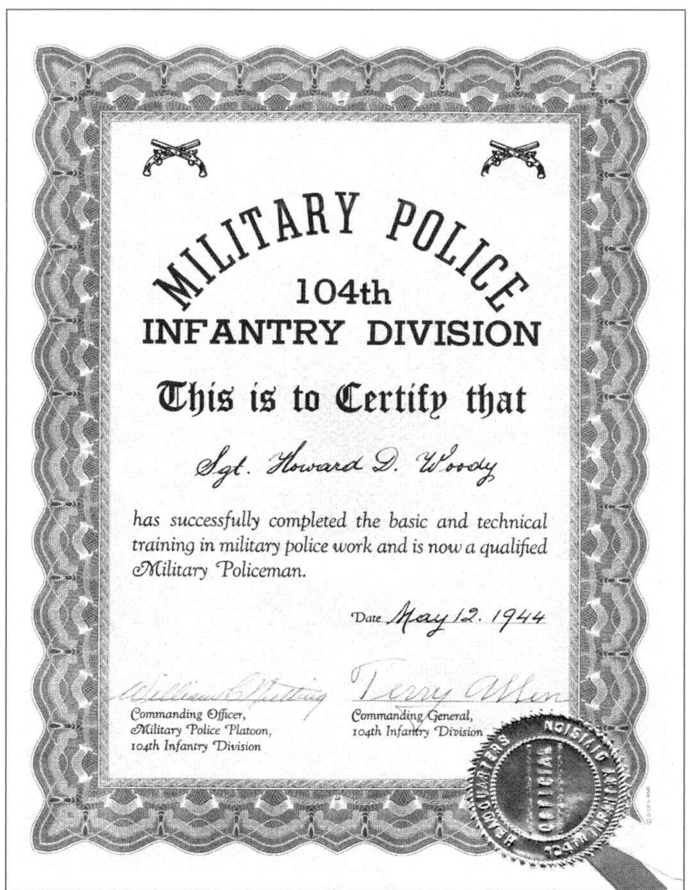

MILITARY POLICE
104th
INFANTRY DIVISION

This is to Certify that

Sgt. Howard D. Woody

has successfully completed the basic and technical training in military police work and is now a qualified Military Policeman.

Date May 12. 1944

Commanding Officer, Military Police Platoon, 104th Infantry Division

Commanding General, 104th Infantry Division

Howard Woody's Military Policeman certification from May of 1944.

When the 104th shipped out, it took a circuitous route through California, Colorado, and Arizona, eventually ending up in New York. Scuttlebutt had it that this route was chosen because the Timberwolves were expected to deploy to North Africa. The rumor maintained that the army brass thought the route would help acclimatize them for their new assignment chasing the remnants of the Afrika Korps across the deserts. Like most US Army rumors, there was no truth in it. The division was destined to serve in Europe, not Africa.

Howard (first from the right) and Esther (fourth from the right) enjoying recreation and relaxation in Oregon as the Timberwolves prepare to ship out.

Howard finally found himself experiencing one of those oceans he wasn't very fond of, as many of the Timberwolves embarked on the luxury ocean liner *Queen Elizabeth.* The giant ship had been converted to a troop transporter, and it spent its war days ferrying troops in both the Atlantic and Pacific theaters. Howard was one of roughly 750,000 troops that the grand vessel delivered to various destinations during the course of the war. The lumbering luxury liner zigzagged across the north Atlantic (trying to avoid Hitler's deadly U-boats) and delivered Howard and his unit to England. However, the Timberwolves were not in the British Isles for long.

Howard recalled landing on the Normandy beachheads between Omaha and Utah beaches. He described the scene in an interview: "We went right up the mountain between Omaha and Utah Beach. It was pretty spooky. The fact is you are so scared that . . . the war got as narrow as one man. The only thing you were thinking about was staying alive. And so you went up through there and after you got through you wondered how you ever did. We got up and out of the shooting and behind the Germans and we got separated from our outfit. We were lost and we didn't have any food."

According to the 104th Division's official history, the remainder of the division landed in France in September of 1944. Howard was a sergeant in the military police platoon for the division, but spent much of his time functioning as a rifleman. The Timberwolves served with distinction in the campaign across Europe, and they saw action in great battles such as Operation Market-Garden, the Battle of the Bulge, and the reduction of the Ruhr Pocket.

The night-fighting skills of the 104th were put to use quickly by their commander, General Terry Allen. The Timberwolves were used in a series of complex operations against German strongholds, such as Weisweiler, Germany, where they infiltrated fortified enemy positions by night and mopped them up by day. War correspondents called them "Allen's smart insomniacs." There are several recorded incidents of men from the 104th infiltrating into German positions by night, and then—once deeply embedded in buildings amongst the German positions—acting as artillery observers and calling in accurate and deadly shelling coordinates for the unsuspecting Germans all around them. The 104th was a division of special men.

Allen was a clever commander. Howard recounted a night operation where General Allen commanded his men to assault a German-held village, but added a special directive. Despite the deep, biting cold of the evening, all the men were ordered to leave their overcoats behind. The Timberwolves thought their commander had flipped his lid for asking them to take off their overcoats on such a bitterly frigid night. The doubts about General Allen's sanity, or lack thereof, rapidly disappeared when the 104th entered the enemy positions and a firefight ensued. The Germans were confused as to who was a friend and who was a foe among the figures that they could barely discern in the dim moonlight. The shivering Timberwolves had no such confusion, nor the resultant hesitation to fire. The men of the 104th promptly shot anyone wearing an overcoat. General Allen was clever in devising this simple technique for his men to easily identify their foes in the pallid light.

Howard was wounded more than once during his service. On one occasion, Howard was hit by "friendly fire" in the form of shrapnel splinters in his neck. Far more terrifying than this impersonal encounter with shrapnel was an instance where Howard was wounded defending himself against a German soldier he encountered while clearing out a building. Howard recalled: "This

German, he was in a hallway. He pulled his rifle up and fired at me. I heard his gun click. He stood there for a minute and I stood there. I just lowered my rifle down. He started running toward me, and then I seen him raise his rifle up. And he had a bayonet on the end of it. I thought, 'Is he gonna stop?' All of a sudden I realized he wasn't gonna stop, so I raised up my carbine and shot him. When he fell he sliced my leg. The point of that damn bayonet went right through my pants and cut a strip off my leg. The scar is still there. I hated to kill him; the only German I ever knew I killed."

Howard never applied for a Purple Heart for either of his wounds. He modestly explained, "Those things weren't that important to me." The memory of his hallway encounter with the German soldier haunts him still, and even seventy years after the event he can still remember the man's face.

These incidents must have made the case of pneumonia he endured seem insignificant, though it easily could have killed him. Howard recalled the incident: "I got pneumonia while I was in Holland. I had to stay behind when the others were ready to move on. A family there took really good care of me. After I got well, I followed the signs that my company had posted on the side of the road telling me where they were headed. They always left signs for those who were separated. The Germans couldn't read them anyway. So, there was little danger. It took me two or three days to catch up to them."

On August 23, 1944, the military police platoon of the 104th received a unit citation for meritorious service. The citation mentions many of the platoon's accomplishments, but gives its most effusive praise to the unit's exceptional performance in keeping the roads open for the Allied counterattack during the Battle of the Bulge. Disguised as US Army MPs, German infiltrators attempted to spread confusion behind the American lines and disrupt an effective response to Hitler's last offensive. This made the situation doubly dangerous for Howard and the rest of the real MPs.

As disturbing as all of these events were, on March 25, 1945, the 104th Infantry Division liberated the Mittelbau Dora Concentration Camp near Nordhausen, Germany. When they arrived, the liberating forces found a scene from hell. Roughly five thousand of the six thousand political prisoners held there and forced to work on Hitler's vengeance rocket program were dead, and the survivors were a mere whisper away from death's long caress. Before fleeing, the Nazi guards had starved, shot, and beaten to death as many inmates as possible. The Nazis left bodies stacked like cordwood, piles of severed arms and legs, and a thousand living skeletons for the Americans to try and save. The 104th must have done this difficult job extremely well, because on the fiftieth anniversary of the camp's liberation, *eight hundred* survivors of Mittelbau Dora attended the commemoration ceremony that was held at the site.

Leon Karalokian, a member of the 104th's Signal Company and an eyewitness to the horrors, recounted:

> Nordhausen was to provide a ghastly traumatic experience never to be forgotten by those who passed through it. On the grounds laid out in neat rows were an estimated thousand decaying corpses ranging from the near skeletonized to the newly dead. In the building bedded on straw side by side in indescribable filth lay the emaciated living, too weak to move away from the dead.
>
> Nordhausen did not have the huge inmate population of Auschwitz. Nor did it have the gas chambers of Dachau or Buchenwald, a refinement apparently reserved for Jewish victims who were on the bottom of the insane Nazi scale of values. Its crematory, though small, was able to convert one hundred of yesterday's men to a heap of ashes each day.
>
> The five thousand victims here, dead and alive, were predominantly Polish slave workers, with some French, conscripted to turn out in enormous underground factories the deadly supersonic V-2 rockets. The full medical resources of the division and from other units did all possible for the living. The very few walking skeletons able to stand wandered about dazedly in the now familiar striped uniforms, bodies shrunken, white faces blood-drained, reaching for the hands of their liberators as they shed their tears. Battle-hardened men of the 104th wept too.

Howard and the other MPs of the division were dispatched to the camp to help save the remaining prisoners, and to oversee the forced recruitment of the Germans in the nearby town to help bury the thousands of rotting corpses. Howard doesn't talk about this episode except to confirm he was there. He

Working in a scene reminiscent of depictions of Hell, Howard snapped this image of rows of corpses at the Mittelbau Dora Concentration Camp.

took a photograph of the camp that depicts the rows of bodies in the process of being policed by the American GIs. The military police assigned to the division worked as a unit to reestablish order in the camp and oversee removal of the piles of corpses. The MPs also forced German civilians from the area to work as laborers, and to witness to the atrocities committed by their nation. These are some of the memories he wishes he could forget.

Not all the events in Europe were as horrifying. Howard once was billeted with a Dutch family for a while, and he was adopted by a nine-year-old girl whose family had converted to Mormonism. They were interested in learning about Salt Lake City and Utah.

Back on campaign, Howard had a chance encounter in the European theater of war with Supreme Allied Commander Dwight D. Eisenhower. The future president of the United States caught Howard shaving off a thick scrub beard preparatory to the general's visit. In a brief chat, Ike learned from Howard that he and his men needed new socks. The Supreme Allied Commander promised that the men would get fresh stockings within a day, and they did. In hindsight, he probably should have asked for a three-day pass to Paris, but clean, dry socks were a true luxury at the front.

Howard and an unidentified soldier sharing a relaxing moment in Germany.

Like so many others, Howard was affected deeply by the war. Faced with the smells of cordite, rotting corpses, and burning towns, he found himself longing for the familiar smell of Wyoming's sagebrush. It was a perfectly sane reaction to the insane world of war. It was a desire to return to a culture of civility, and to leave the memories of the limits of human cruelty behind. Esther, the love of his life, and his first son, John G., were so far removed from the horrors of the war in France and Germany. Howard's old life in the Rocky Mountain West must have seemed like a hazy dream calling to him in those quiet moments. Even seventy years after the war, Howard still avoided talking about it if he could help it. His reluctance to recall his acquaintance with the restless ghosts of France, Holland, and Germany was not difficult to understand.

KEEP THE HOME FIRES BURNING

When Howard left for the battlefields of Europe, Esther stayed in Mountain View. Bonnielee recalled that Esther had a room in the telephone office until just about two months before her baby was due. She further recollected that at this time, Esther was bunking in a double bed with Lottiebelle Woody, with whom she shared operator duties. Esther was moved into a cot in the kitchen with John D. and Mary Ellen. Howard recalled these events differently, and indicated that in his recollections Esther lived in the small house, and that Dr. McDill delivered her baby, John G. Woody, there in February of 1944.

Esther tried to keep in touch with Howard during his deployment by writing a letter every week. Only a few of these letters reached Howard in the field. Like many of his fellow dogfaces at the front, Howard was not good at writing letters to loved ones at home.

After Howard rejoined his unit, baby John filled Esther's life. She took baby John into the field when she was driving for John D., who—even in his sixties—still continued to repair downed lines during the war years. Esther cared deeply for John D. and did not wish him to be in remote areas alone, as it increased the potential risks during repair and maintenance work. Repair calls often meant going out in the deep cold, blinding snow, and slashing wind of the Wyoming outback. Esther kept John G. by her side, wrapped in a blanket in the truck, while John D. fixed the lines. Baby John G. was her direct link to Howard during these dark days of separation. Her son was yet another Woody literally born into the family business.

As an infant, John G. had the reputation of being a handful. For example, the toddler was fond of getting up and running on the kitchen table. When Esther had to leave the kitchen to use the outhouse, she attempted to thwart this behavior by moving all the chairs to an end of the kitchen away from the table. By the time she returned, the energetic John G. had moved a chair across the room and was gleefully running atop the table. One time, John G. took advantage of Esther's brief absence to abscond with her pet magpie. After a brief search she found the flustered bird in a jar, where John G. had hidden the poor animal.

When money was tight, Esther turned to sewing to help add to the family's income. She sewed custom-made dresses for her customers, and, according to John G., she had exceptional skills as a seamstress. On her sewing machine, Esther produced much of what the family wore. As with everything Esther did in her life, she threw herself into the task and mastered it.

WHEN HOWARD COMES MARCHING HOME AGAIN

With the end of the war, Howard and millions of other service men and women returned home. It was a sweet time for the young man from Mountain View. Howard talked about the joy of getting on a troopship headed for home. His friend, Garvis Richie, was on the ship with him. They smiled broadly, realizing they had survived the war. Howard recalled of his pal Garvis: "I used to laugh at him as he always would smoke cigarettes and pray. I didn't smoke cigarettes and I didn't pray. The pray thing was all because of this Mountain Meadows Massacre and my grandfather being the only survivor that we knew."

As relieving as it was to take ship and sail to America, the best memories concerning Howard's return were created as he once again filled his senses with the familiar sights, sounds, and smells of his beloved Wyoming. In his own words, Howard recounted: "The first thing I did, as I remember, I pulled off on the Bigelow Bench, and it had been raining just a little. I could smell that sagebrush! I pulled right off and rode down into that sagebrush, and got out and got some sagebrush, just so I could smell it. It sure smelled good to me. It smelled so good! I just stopped there and sat on the fender of the car and looked out on that sagebrush on the Bigelow and thought how wonderful it was to be back."

Howard was finally reunited with his beloved Esther and their son, and life was good again. When recalling Esther, Howard's face always lights up. "I had the best wife any man ever had. She could do anything. She was a good carpenter. She helped me build a house. We built that house from scratch. To build that house we borrowed twelve hundred dollars. We cut the timber, cut the logs, hauled them down, had them sawed up, got them dried, and did everything we could to save money, because we didn't have any. But I'll tell you, she was the greatest thing in my life. She was just somebody to look up to all the time." Howard remembers his days of working on the house that he and Esther built as some of the happiest in his life.

Once out of the military, Howard began to seek employment to support his family. The cattle were all gone from the ranch, and it would take time to rebuild the herd. Howard helped his father with the telephone company, but he also looked for outside employment.

I'VE BEEN WORKING ON THE RAILROAD

Between October of 1945 and April of 1947, Howard took a job with the railroad working on a survey party. His job was as a rod man, the individual who holds a graduated rod so that the surveyor can measure the height of objects at a distance relative to the survey instrument. Howard often provided his own horse and saddle to get to the isolated projects.

The crew was attempting to survey a proposed tunnel through the mountainous terrain near Altamont, Wyoming, roughly forty-five miles west of

Mountain View by air. In 1901, the Union Pacific Railroad constructed the roughly 1.5-mile-long Aspen Tunnel through Aspen Mountain. This single-track tunnel eliminated the long, winding grade going over the top of the mountain. As traffic increased, Union Pacific Railroad began to look for a location for a second tunnel, called the Altamont tunnel. Howard worked on the surveys for this second tunnel, which was not constructed until 1949.

In 1947, heavy snows threatened to terminate continued survey work for the remainder of the season. The survey crew was waiting in the railroad depot while the foreman, Bob Brown, talked on the telephone to his boss in Omaha. The foreman informed his distant boss that the snows were already too deep, and that winter in Wyoming was just starting. Standing nearby, Howard couldn't help but hear as the two decided to put off the survey until spring.

When the foreman finished the call, Howard said, "I think we can do it, if you want to." The foreman looked over this lanky young upstart, apparently sizing him up, and asked Howard how it could be done. Howard spoke right up about how people in Wyoming had been getting such things done for years before the railroad came to Wyoming, and about how horses could get through the deep drifted snows when motorized vehicles could not. As an accomplished rider, Howard understood that it was still possible, with some difficulty, to reach the areas to be surveyed. As a born Wyomingite, he knew that the only way to avoid being beaten by the Wyoming winter was to face it head on. He proposed filling one half of a railroad car with hay, and the rest of the car with pack animals, Howard's horses, equipment, and saddles, and moving this loaded car to Altamont. From that mobile base they might stage the project, which would be accessed by surveyors on horseback.

The foreman thought over Howard's bold proposal and decided to support it. The boss in Omaha agreed to the gamble, and soon Howard was out in the deep snows. He recalled, "It was a hell of a battle, going up over that mountain." Despite the bitter Wyoming weather, they finished the job. The chief engineer, John Perkins, reviewed Howard's notes at the completion of the survey and pronounced them to be deserving of a reward.

Howard had time while working on the project to think and reflect. His thoughts were filled with his father's struggles to build the telephone company back in Mountain View. With the railroad survey completed and accepted, Howard made an announcement: "I gotta go home and build a telephone company." The foreman, who had been thrilled with Howard's efforts on the project, threw a fit. His mood calmed a bit when Howard explained that his father was getting old and needed his help.

Despite the railroad's tremendous offers and incentives to stay (including putting him through engineering school with the guarantee of a job when he finished), Howard stuck to his decision. He had a family to provide for, and though the offers were tempting, there was only one path for him to follow. It was a difficult choice for the young man, but he turned his steps toward his father's dream, not realizing it was his own destiny that called.

Howard had grown up around the company he seemed destined to lead. There were long field trips with his father to work on lines, and although John D. asked the young Howard to help him when he could, he did not directly pressure his son to prepare to take over the company. As a child, Howard did not trouble himself with thoughts of following in his father's footsteps; rather, he spent much of his time working on the ranch. He loved the freedom of being out with the animals and not having anyone to answer to. It was on the range he always felt most free. It was a pattern that stayed with Howard all of his life. The ranch was that place where he could clear his mind and renew his spirit.

At that time, John D. was in his sixties, and his ability to meet the strenuous physical requirements of the work was diminishing. Howard divided his time between the ranch and Union Telephone Company. Since John D. had sold off

MS. INFORMATION

One night a customer called in and asked for the number of a man by name. The operator looked for a listing but could not find anything under the name given by the caller. She said, "I just can't find his name, does he have a wife?" The indignant caller replied: "Yeah, he's got mine! That's why I'm looking for him!"

the family's cattle during the war, Howard set about reestablishing the family's cattle herd, a task that took roughly thirty years. With John D. only capable of taking on less and less of the backbreaking fieldwork, Howard eventually took over all of the physical duties at the telephone company for his aging father.

John D. and Howard cut poles, processed them, dug holes, set the poles, and then hung the heavy cables. They went out in storms and fixed downed wires and installed new lines. They hooked up new customers and serviced equipment. Like John D. Woody, Howard nimbly climbed up the poles using his hands and the linemen's spurs that were strapped to his legs. The ranch continued to provide sustenance for the family. The entire family, including Howard's sisters Lottie and Bonnielee, his wife, and even his mother continued their commitment to the telephone company and its customers. The family worked as if with a single purpose: to keep John D. Woody's dream alive.

Howard and John D. sometimes had difficulty getting the board of directors to act in the best interests of the company. In 1947, out of frustration with the board of directors, Howard assembled an advisory board of experts from the telephone industry in an attempt to sway Union's board as to a course of action—but to no avail. Howard told the story thusly: "I remember one time there was a little telephone company up in Farson. They decided they wanted to sell out and wanted somebody else to do it because they weren't doing a very good job. So I brought it up to the board of directors. I had one old knot-headed director, Harry Buckley. He said, 'No, by God, what do we want to go up there and spend our money for?'" Following Harry's lead, the board rejected the acquisition. Howard continued: "I just sat there. Then I said, 'Gentlemen you will live to regret this decision!' And they did." It turned out that Farson, Wyoming, was positioned as a key lynchpin between Rock Springs and Jackson Hole. Farson was also the gateway to South Pass, the great gate of the Rockies. Farson was positioned advantageously to extend service to the future gas fields of southwestern Wyoming in the vicinity of Pinedale and Granger. Howard did not know about the coming energy booms, but the strategic location of Farson—a key crossroad linking many important Wyoming communities together—made it prime territory for expanding the Union Telephone system.

The father and son team also found that they were having difficulty getting the board of directors to approve expenditures necessary to run and expand the company. As Howard put it, "They didn't know anything about the operation side." As a solution, Howard and John D. proposed leasing the company they had built. The board members, including Cliff Anderson, George Smith, and Harry Buckley, agreed to the Woody's lease proposal, provided they put all the money they made—minus their expenses—back into the company. This untied John D. and Howard's hands on the day-to-day decisions necessary to run the company efficiently. The larger issues, such as obtaining loans and expanding service coverage, remained in the hands of the board of directors. The leases did not list Howard, but were in the names of John D. and Mary E. Woody. These were periodically renewed, the final lease being issued in 1956 for the cost of one dollar.

John D. Woody's health continued to decline throughout this period. Howard recalled: "Now dad took me to walk the lines for him instead of the other way around. In the early days, he [John D.] rode the lines on horseback to make sure there weren't any lines down. People wouldn't always tell you if their phone didn't work, so if the operators couldn't get your line connected for a call or unless someone complained, you never knew if the lines were up or not. Now he relied upon me a lot."

I'M GONNA MAKE THEM AN OFFER THEY CAN'T REFUSE

Howard had made his decision. His father had worked hard during the war years and beyond to keep the company functioning. John D. was growing old, and Union Telephone remained in poor financial shape. It was time to decide what to do with the company. Howard and John D. knew that they could not continue to lose money on the company forever. John D. had poured his heart and soul into the company, but it seemed like the time had come to retire and sell his dream. John D. had built a good telephone system, and maybe someone else could turn a profit with it.

In 1948, father and son met with representatives of AT&T and offered them the company for one dollar. As part of the deal, Howard agreed to continue servicing the local lines. According to Howard, AT&T turned down the deal and followed the curt refusal with a statement to the effect that Union's customers were not worth the time of the telephone giant. Howard and John D. were angered by the answer they received. Sometimes such a dismissive pronouncement is all a person needs to find that extra something inside that drives them to take their efforts to the next level. In this case, it was a blatant reminder of why John D. had built Union Telephone in 1914. That simple statement by the AT&T representatives about the insignificance of Union Telephone's corner of the American Outback was a driving force in remaking the company.

Howard threw himself into the telephone company with renewed vigor. Ever since John D. and Mary Ellen had leased Union Telephone Company from the board, the family had been buying back stock when they could. Their ultimate goal at that time was to acquire the controlling interest in the company. Even so, they were still willing to let stock go, as circumstances warranted, if it helped with the expansion of the company. Howard used Union Telephone stock much like his father—as capital with which to build the company. Examples of this occurred when Union swapped stock with other exchanges and managed to acquire the Bridger Telephone Company in 1946, and then the Lyman Telephone Company in 1956.

Howard recalled their campaign to buy back the stock that was not in family hands. "Dad had never owned controlling interest in the company during his time he operated the company. . . . Because dad had offered stock for payment so often, it took years and a lot of money to buy back the stock." Howard's daughter Bonnie Jean Shannon added, "Mom threatened to paper the walls of the house with stock, because there was no money for the house."

HAPPY LANDINGS

After reading an issue of *Popular Mechanics,* Howard devised a unique solution for getting around in the deep snows of winter. Inspired by the possibilities of rapid travel across snowy landscapes, he converted an old airplane into a snow plane.

The engine for the plane had come from a plane crash. In 1948, Howard was in the field near Nebraska Flats with employee and future brother-in-law Ray Tanner when they heard the noise of a plane in distress.[2] The plane struggled against the icy winds, but it was clear to Howard and Ray that the plane was in trouble. Howard said to Ray: "That plane won't make it. It's not got enough power." The two watched anxiously as the plane lost altitude and crashed into the frozen landscape.

In their tracked vehicle, Howard and Ray rushed to the crash site thinking the pilot must be dead, but luckily he was only a little banged up. Ray and Howard gave the man a ride to safety. Later on, Howard inspected the crash site, and, believing the airplane's engine was serviceable, he offered to buy it from the owners. There was a little haggling about price, but finally the owners of the wreckage took Howard's offer. The damaged plane and engine became the nucleus for the new snow machine, which was put together the following year. The wings of the plane were left off, and three runners replaced the landing wheels. The propeller was a push type, and the engine was rated at sixty-five horsepower. The rudder was left in place for steering. The noise in the open cockpit was deafening, but she was a fast little vehicle.

Howard and Ray cut a hole in each of the skis and installed an ingenious pair of spring-loaded spurs that could be deployed by a trigger in the cockpit. When the switch was activated, the spurs sprang through the holes in the skis and created points of drag to help slow the plane. It was the closest thing the snow plane had to brakes. In the right conditions, the spurs were *almost* like brakes. The retractable spurs slowed the machine, but stopping the machine on a dime was out of the question.

At the time the snow plane was assembled, modern snow machines were more than a decade in the future. Snow machines of various types had been in service since roughly 1910. Howard's snow plane was capable of moving two people and equipment across the frozen landscape in relative comfort. With an open cockpit, the warmth of the passengers depended upon how well their cold-weather clothing functioned. The plane became something of a local legend.

[2] *Ray married Esther's sister, Mary Sue Gladwill.*

One can imagine this powerful machine speeding across the rugged, snow-covered terrain of Wyoming. The engine roared loudly through the otherwise silent winter landscape, and the propeller kicked up a great plume of snow in the vehicle's wake. The occupants of the winter ranges, such as elk and antelope, were startled and confounded as the whining machine filled their world with strange sights, sounds, and smells. This frightening contraption was a forerunner of the thousands of snow machines to come.

The snow plane was not the first instance of Howard using some of his innate ingenuity to create a strange vehicle. His sister Bonnielee recalled that as a youth, Howard once "borrowed" the engine from the family's washing machine and married it to a small wagon. Bonnielee was thrilled that her brother gave her rides through town with his homemade automobile. Their mother, Mary Ellen, was less than amused, and Howard's first experiment with alternative transportation was quickly brought to an end. In short order, the engine was returned to its proper home and was once again washing the family's laundry.

The snow plane was used to check all of the lines, as well as ferry needed supplies and mail during the infamous winter of 1949. During that winter, much of Wyoming, South Dakota, Nebraska, and Colorado were covered by monster blizzards that produced fifteen-foot-plus drifts in many places. Ray Tanner recollected drifts up to fourteen feet deep near Urie, and others reported drifts as deep as twenty feet. That winter became the local standard for judging the severity of winter weather, replacing the infamous winter of 1886 for that dubious distinction.

The snows of 1949 were too deep even for the mighty diesel driven passenger trains to plow through. Unable to continue until the tracks were cleared, the railroad stranded hundreds of travelers at isolated depots. Both the population of the area and all of the travelers passing through were snowbound until help could get to them. Communities such as Green River, Evanston, and Rawlins did their best to assist the stranded travelers, providing food and shelter for hundreds of unexpected guests.

Along the Lincoln Highway (US Route 30), automobiles and trucks were buried in the deep snows, with passengers sheltering inside. The crude snow buggies of the period hurried pregnant women to hospitals, as emergency responders scrambled to deal with the unprecedented crisis. Jim Woody was born five days after the first of the monster blizzards hit, but there is no indication that his birth involved any more drama than is usual in such circumstances. Howard was kept busy hauling supplies and mail to his neighbors. The snow plane also served as a taxi service, ferrying individuals to areas where roads had sufficiently been cleared so that they could use conventional forms of transportation.

During the height of the crisis, the military airlifted supplies to humans and livestock alike. The Air Force was sometimes a little ambitious with the hay bales it airdropped, occasionally hitting the cows they hoped to save, and at least once almost crushing a rancher with a plummeting hay bale while the unsuspecting man was checking his animals. In an emergency such as this, maintaining a working telephone system was critical to ensuring that the

Howard and Ray Tanner built the snow plane using parts from a crashed airplane.

rural families were safe. It was too easy for isolated occupants of the region to be forgotten at such times. The snow plane gave excellent service under the harshest conditions during its maiden year. For roughly a month-long period of the bitterest winter conditions in Wyoming's recorded history, Howard was piloting the snow plane every day, getting assistance to his neighbors.

On another occasion, the snow plane was used to rescue grocer Hal Benedict and ranger Bert Clark, who were skiing up near Bridger Lake. The pair had traveled roughly twenty miles from their starting point, and things were going wrong. Hal's feet were covered in painful blisters, and Burt was having difficulties with the cold. The skiers sought shelter in a ranger station, which was buried deeply in snow. Since the door and windows were buried, Hal had to climb atop the roof and enter the building through a loft window. Despite finally being within shelter, Bert was slipping into hypothermia and was in danger of freezing to death. Hal struggled to warm the hypothermic ranger, and with some effort finally managing to stabilize the suffering man. However, both men knew that neither of them was in condition to ski to safety the next day.

The ranger station had a working telephone line, and the call for help went to the Union Telephone office. Mary Ellen took the call, and Howard responded by taking the snow plane on a rescue mission. At full throttle it took Howard an hour to get the snow plane up to the cabin site. When he got there, only one of the two men could fit in the snow plane. Bert told Howard, "I'll let you tow me. Hal can take the cab seat." With its new charges in tow and in the rear seat, the return trip of the snow plane was much slower that the trip to Bridger Lake. Bert had to hang on to a line behind the push engine on the craft. He was buffeted by the prop wash of the plane, plunging him into a localized vortex of deep wind chills. Bert had great difficulty remaining upright and fell several times, forcing Howard to deploy the ice spurs and eventually bring the snow plane to a halt. At about the halfway point in their journey, Hal and Bert changed places. Howard brought both of his charges in to safety. With some pride, he recalled, "I tell you, them guys were awfully thankful."

MOONLIGHTING WYOMING STYLE

Howard sometimes held jobs outside of the company to support his family. He worked for Bridger Valley Electric Association (BVEA) for roughly a decade, starting a long relationship with that cooperative association. Howard served on the board of directors for that company and maintained a good relationship with BVEA long after his employment ended. In the late 1940s and into the 1950s, Howard served as the supervisory lineman for BVEA. Typically, he first worked his eight hours for BVEA. After putting in this wage-earning shift, Howard worked on the Union Telephone system, often into the wee hours of the morning. There were all-too-frequent twelve-to-sixteen-hour workdays in Howard's life. Howard just shrugged off the long hours, and he recounted, "The moon seemed to be out longer and brighter when I worked nights; more than I had ever noticed before."

Howard spent years working for Bridger Valley Electric as a lineman, a job that permitted him to continue working for Union Telephone during his "off hours."

In 1953, one of Wyoming's infamous spring snowstorms pelted the area with heavy slush and snow. As the storm progressed, ice thickened on the lines and the supporting poles, the heavy weight of which eventually collapsed both. Most of Union Telephone's lines were disabled by the storm. Howard was working for BVEA when the storm hit. BVEA's system was equally affected, with much of the power system being inoperative. Suddenly, Howard found his usual back-breaking twelve-plus-hour days were inadequate to deal with the emergency. It is not clear how he managed, but Howard put in over seventy-two hours straight working on the two systems during the emergency. Howard had help from an able crew while repairing the BVEA lines, including Jack Byrne, Wayne Herdsman, Blain Meeks, Bill Riding, and Tony Riding. When Howard ended his long days with BVEA, he had only John D. to assist him with the heavy work for Union Telephone. Lines were down all over, and one of them had to be restrung across the Green River and—once across—up a mountain slope.

FAMILY MATTERS

Love is like a dyin' ember
Only memories remain
Through the ages I'll remember
Blue eyes cryin' in the rain.
—from *Blue Eyes Cryin' in the Rain* by Fred Rose[3]

Howard and Esther did not have a picture-book marriage. Instead, it was one of those pairings most people dream of—a genuine marriage. Esther jokingly told people: "Howard is married to Union Telephone and it just happens I'm his mistress. The mistress gets better presents." Esther and Howard did not completely share in their likes and dislikes. For example, Howard loved to dance, but gave it up once informed that his new bride did not share his terpsichorean passions. The couple spent long hours enjoying Walker Lake, one of their favorite spots. Ultimately, they shared the most important thing: an abiding love and respect for one another.

[3] *This song is the one that most reminds Howard of Esther.*

The Woody family continued to grow. Esther and Howard's first son, John G., was followed by Gloria Dee (1947), James H. (1949), Lee J. (1952), and Bonnie Jean (1958). None of the children were specifically groomed to run the family business. However, the ranch became a place where they learned responsibility, hard work, and making the best of what the climate lets you have. It was the crucible that helped to build the ethics that Howard and Esther wanted them to apply throughout their lives. Each member of the third generation made their personal decisions as to whether they wanted to have a career at Union Telephone Company. This was much the pattern that Howard apprenticed under, and it has become part of the model for bringing family members into the company. It allowed the youngsters to develop a work ethic and learn the importance of the family members coming together for the common good. The ranch had a way of teaching a person to pay attention to nature, be creative in solving problems, get used to hard work, and develop a sense of practicality. It was the perfect place for bonding the family and passing on Woody family traditions. The lessons learned at the ranch followed the family members throughout their lives.

Esther teases Howard with a piece of their fiftieth-wedding-anniversary cake.

While they were growing up, the lives of the third-generation children were very much like those of rural children throughout Wyoming. The youngsters had to make their own entertainment, and oftentimes they devised unusual diversions. According to generation-three member Bonnie Jean Shannon, the kids used to drive their Great Grandma Katherine Gladwill almost to the point of pitching a fit by licking the salt blocks left in the field for the livestock. Bonnie further described the popular pastime of teasing tarantulas by tricking them into thinking they had prey in their tunnels. Other forms of entertainment included drowning ground squirrels, beheading grasshoppers, and breaking up mud-dauber nests.

In time, the children were given jobs such as cleaning telephones.[4] They were gently led to test themselves at working in the family business. Son Jim remembers that each of the third-generation children were paid a dime each to strip the bark from poles, preparatory to treating the wood with creosote. This turned into a lucrative activity for the youngsters during the summers off from junior and senior high school.

If Howard was the captain of the ship, Esther was the admiral of the family. Howard took a more hands-off approach to raising the family, and he was often seen tending the livestock during his hours away from telephone and electrical work. Esther took care of the children, was an avid painter, and, in her spare time, continued to help Howard with the company. Though a petite woman, Esther was a force to be reckoned with. Descriptions by those who knew her use terms like "firecracker," "spitfire," "pistol," and "strict but fair." She was a lot of woman in a small package, and she was one of those unforgettable characters that breathed life into all she encountered. Esther spent hours on her paintings and photography. She learned how to do carpentry, and—according to many family members—her chocolate chip cookies were to die for.

One thing was certain: a person knew where she or he stood with Esther. If Esther had an opinion about something or someone, she made it crystal

[4] *At that time, telephones were owned by the phone company and were rented to customers. When telephones were turned in, they needed to be cleaned before being put back into service.*

clear. Esther was not one to mince around the heart of a matter. She could be a stickler for rules, but she was always compassionate about her children and employees, even when teaching a hard lesson. She had what entertainers refer to as "stage presence," that ability to draw the attention of the audience even when not delivering lines.

Esther was the perfect foil for Howard. Her husband was strong, levelheaded, and shrewd. Howard enjoyed his solitude, and his loss of hearing (a result of the war and driving the snow plane) perhaps made him a little less reachable than he might have been before his service. He and Esther shared a strong work ethic and belief in family and community. For a man with a sweet tooth, having a mate who was a gem at baking was a real plus. Esther and Howard adored each other, and they allowed one another the space and encouragement to grow as people.

One of Howard's regrets in life is that he did not have more time to spend with his children while they were growing up. To meet the needs of his multiple jobs, he was often out the door before they awoke, and he returned home after they were in bed. He lowers his head and almost whispers as he gives voice to his regrets. "I should've just taken more time anyway. But I didn't do it. We was just so damn poor. We needed to use every minute to get to work and make money."

UNION EXTENDED FAMILY

For many people in the company, Esther became a mother figure. Esther did occasional shifts as an operator, and she was instrumental in making sure that the operators had full employee benefits, despite being part-time employees. When employees tried to save the company money by keeping the windows open on hot days, Esther told them to close the windows and use the cooling system. She truly cared about the employees, and she took that attitude with her as she helped shape the company's Human Resources program.

In Esther's heart and mind, the employees were part of her extended family. She arranged social events—such as picnics and parties—to express appreciation to all those who worked to make the company a success. The company held its first Christmas banquet in 1960, at which everyone received a frozen turkey from the

Esther and Howard enjoying each other's company.

company. Esther never sat on the board of directors, but make no mistake about it—she affected the decisions of the leadership. She served for a time as the director of the Human Resources department, and she was instrumental in making Union Telephone a cordial working environment. She also served on the management team. Most of all, she was Howard's muse and partner through good times and bad.

The Woody family recognized that Union Telephone owed its success and longevity to its loyal, hard-working employees. It was their labor that kept John D. Woody's dream alive. They have endured the tough times, the long hours, and the difficult weather to help carve out a place for Union in America's Outback. Howard and Esther made sure the employees were paid well and taken care of. Howard summed it up: "You know I spent most of my life worrying about employees. I didn't want to be the same kind of son-of-a-bitch that I had worked for in the past."

ANY PORT IN A STORM

Wyoming was infamous for its dangerous driving conditions during the winter months. Many places in Wyoming had refuge structures or caches built where stranded travelers might find shelter, heat, and food while waiting to be rescued. Some of these caches even had fodder for the livestock of those who were on horseback or otherwise traveling with livestock. Union Telephone Company installed a halfway station with emergency telephone service between Lonetree and Mountain View. The cache was used to good effect on several occasions by travelers in distress. The construction of this refuge was yet another example of how Union Telephone looked out for their neighbors. There were no tax breaks or marketing incentives to build a cache at a remote roadside. It just made sense to build this oasis for the times when their neighbors might be in need of a helping hand. That time eventually arrived, and the cache helped save travelers in distress.

BROTHER CAN YOU SPARE A DIME?

You can't borrow yourself rich.
—*Attributed to John D. Woody*

Union Telephone's continuing track record of unprofitability convinced Howard that he needed to change how the company did business. The company mainly served small pockets of rural subscribers. Howard wanted to expand the company sufficiently to add a significant number of customers. There were additional potential customers scattered amongst the mountains and valleys of America's Outback. The addition of the Lyman exchange in 1956 helped expand the number of customers, but Union Telephone needed many more to get to profitability.

To accomplish the expansion envisioned by Howard required capital. It was not certain that the banks would support such a risky venture. Bankers of that time period thought much like the AT&T board: they had little interest in servicing scattered frontier populations. However, there was the Rural Electrification Administration (REA), an organization that was able to give Union Telephone just the kind of help it needed. The REA grew out of the realization that rural Americans likely could not get access to dependable electrical or telephone services unless the federal government supported those who were building such improvements. The REA was authorized under the Emergency Relief Appropriation Act of 1935 (and other acts) to make low-interest loans to corporations providing electrical and/or telephone services in rural America. In a 1973 amendment to the Rural Electrification Act, Congress declared that "rural electric and telephone systems should be encouraged and assisted to develop their resources and ability to achieve financial strength needed to enable them to satisfy their credit needs from their own financial organizations and other sources."

Howard first started his campaign to approach the REA for a loan in 1953, hoping to borrow money for converting Union Telephone to a dial-tone system. Initially, John D. and Howard did not agree about approaching the Federal Government for assistance on a loan, but Howard was convincing, and they decided to take the risk. It was not John D. that held back the application; the Union Telephone board of directors had to approve the application for such a loan. The reason given by the board for not going forward was dumbfounding to Howard and John D., and it demonstrates how out of touch the board was with the operations of the company. Howard described the situation: "I worked nights getting all those maps ready, and got it all done. Sent it to Washington, got approval, and old George Smith said, 'Cows are a bad price this fall, and we ain't going to do it.' And I thought the dumb old son of a gun didn't know what he was doing."

Amos Jackson, a consulting engineer, was the first non-operator employee to work for Union Telephone. He was hired to coordinate the engineering demands required for the Flaming Gorge project.

George was not bluffing regarding his curt dismissal of the expansion. Under George's leadership, the board of directors voted down the loan application. It was not the first time in Wyoming the price of cattle affected a business decision, but it remains baffling that the commodity market was affecting a decision about expanding and modernizing a utility. After all, Union Telephone did not own a single cow.

Howard received the board's approval to apply for the loan again in 1956. As part of this process, Union Telephone was required to approach the FCC and make an application for a Certificate of Convenience and Necessity, a document that gave a company rights to service an area. The company was required to have projects engineered. Union Telephone responded to the requirement by hiring an engineer. One of Union Telephone's first non-operator employees was consulting engineer Amos Jackson of the company Jackson and Jackson out of Salt Lake City.

The board of directors that approved the loan had been expanded from three members to five, and it included Howard (the first time a Woody family member served on the board), Cliff Anderson, Harry Buckley, Ernest Dahlquist, and Albert Neff. Howard was appointed manager of the company in June of 1956. It was at this time the company name was changed from Union Telephone of Mountain View to Union Telephone of Wyoming. Nationwide, there were several telephone companies named Union Telephone, and the change was requested by the REA.

This is the Dutch John central office and microwave dish. It was installed to service the Flaming Gorge Dam workers at the man camp in the new town of Dutch John.

It took months for the REA to process Union's application, but in the end the efforts were rewarded. In the late 1950s, Union Telephone received the first of several REA or REA-guaranteed loans. Initially, it appeared that the REA was not going to finance Union's proposal, but Howard eventually managed to win his loan.

The first REA loan was for $300,000, and it was granted to Union Telephone in December of 1956. The REA loan allowed Union Telephone Company to receive the capital it needed to rebuild the exchanges and help install the dial-switching equipment. At this time, the company expanded the system to the Flaming Gorge dam-project area, and the loan helped buy necessary equipment. For Union Telephone's bottom line, the loan (combined with the expansion to Dutch John and Union's acquisition of the Lyman exchange) turned out to be a game changer. The REA provided the needed capital for both expansion and the upgrading of equipment. Howard had successfully positioned the company as a real competitor in the region.

It was not until 1956 that Howard was officially made a salaried manager in the company. If you count the years he helped his father inspect lines, Howard had worked for the company—in one capacity or another—for roughly twenty-four years before this milestone was reached.

Prior to the absorption of the Lyman, Wyoming, and Summit County, Utah, telephone franchises in 1957, Union Telephone provided service to the communities of Manila, Linwood, Burnt Fork, McKinnon, Lonetree, Robertson, Fort Bridger, and Mountain View. There also were unincorporated areas such as the Henry's Fork of the Green River. The expanded Union Telephone service locations encompassed five counties in three states.

The year 1957 saw frenetic action as the company expanded and improved facilities. Union connected to AT&T's B Cable near Urie, Wyoming, providing expanded long-distance services. In April, the company ordered new switchboards from Leich Electric for Manila and Mountain View. By June of 1957, Union Telephone was letting contracts to build an office building at Dutch John, and a power line from Dutch John to Manila. The company also began its search for land for central office facilities in Mountain View and Manila. Things were starting to move much faster. By August of that year, these building sites had been acquired, and the construction contracts for the new central offices had been awarded.

Some years after the events of 1956 and 1957, Howard recalled a conversation with his father on the topic of the loan. The two were sitting out on the porch watching a beautiful Wyoming sunrise. The smell of breakfast being cooked in the kitchen mixed with that of the damp sagebrush. John D. turned to Howard and said, "About borrowing that money, I think you done the right thing." It was one of Howard's proudest moments. Typically, he understates the importance of the conversation: "That was hard for him to say. It made me feel pretty good."

FLAMING GORGE DAM

In the mid-1950s, the United States Bureau of Reclamation initiated a hydroelectric project that impounded water behind a dam in Utah near its border with Wyoming. The project would create a reservoir stretching ninety-one miles upstream, deep into the Cowboy State. The reservoir was part of the Colorado River Storage Project, and it is among the largest in the West. The project was started in 1956, and completed late in 1962. This huge undertaking injected a massive amount of cash into the local economy. The dam itself cost $49,600,000 to construct. The power plant (annually generating roughly 500,000 megawatts of power) and its associated switchyard cost another $65,000,000.

The completion of a dam in Flaming Gorge was a long and complex process that started decades before construction ever began. As far back as 1919, Utah Power and Light Company had considered a dam on the Green River for hydroelectric power generation. However, the complex issue of equitable ownership of water in the West moved the project off the drawing tables.

In 1949, the states of Wyoming, New Mexico, Colorado, and Utah agreed to apportionments of water in the upper Colorado drainage system under the 1922 Colorado River Compact.

In 1949, three sites were considered for construction: Echo Park, Horseshoe Canyon, and Red Canyon. The Echo Park locale ran into stiff resistance by environmentalists, so the Red Canyon site was chosen as a sacrifice to protect Echo Park.[5] The dam at Red Canyon backs the water up through Flaming Gorge, known for its stunning palisades that comprise red, yellow, gold, and orange sandstones that glow like a warm fire, especially at daybreak and sunset. The dam was designed to both impound water and manage the flow of the river through Lees Ferry, but it also included a major hydroelectric component. The Green River was expected to earn its keep in more ways than one.

Union Telephone Company almost lost the Flaming Gorge service contracts to its competitor, AT&T/Mountain States Telephone, which tried to discourage Union from providing telephone and Teletype services for the project. Initially, The Bureau of Reclamation informed Union that their Flaming Gorge project would likely not need service until 1957. This gave Union plenty of time in 1956 to complete their REA loan application, or so they believed at the time. Mountain States Telephone stepped in and sought to pull the project out from under Union Telephone. Mountain States had arranged a meeting with the Bureau of Reclamation for the service, and they asked Union Telephone to step out of the way, claiming that Union Telephone would not be able to secure the funding in time to provide the Bureau of Reclamation the services they desired.

The Union Telephone representatives listened to the Mountain States representative, Charlie Parker, drone on about the project. Finally, Howard interrupted. "Charlie, I have listened to you speak for two hours about how Mountain States will benefit from providing service to Flaming Gorge. What I haven't heard is how our withdrawal will benefit Union Telephone." Charlie appar-

[5] *In the early 1960s, another component of the Colorado River Storage Project, the Fontenelle Dam, was constructed miles upstream near La Barge, Wyoming. Combined with the Flaming Gorge reservoir, these two sites impound sufficient water such that there are recurring proposals to build huge pipelines to take Wyoming's water to the Denver metro-complex.*

ently replied, "Christmas only comes once a year." Howard grinned as he retorted, "You're right Charlie, and this year we aren't giving you a Gawd dang thing!"

After this meeting, Union contacted Senator Frank Barrett's office to garner his assistance. The letter to the Senator stated:

> If we could have the same agreement with the Bureau of Reclamation that Mountain States now has, then we could go to a banker or private investor and obtain the necessary funds to get the service into Flaming Gorge by the time the Bureau desires it, and then let the REA loan take its regular procedure. It all ties in with our plans, for we have the same line and facilities to build regardless....
>
> Union Telephone is a small company and it is not financially able to carry out a long legal entanglement with Mountain States at this time, and yet I am confident we are only wanting what is rightfully ours. Flaming Gorge is rightfully ours to serve and losing it would even perhaps be denying this area the telephone service which it is entitled to.
>
> Mr. Stone of the REA seems to be very inconsistent and contradictory in his advice to us. I wonder what the REA's policy in this matter really would be; last week he told us we could certainly get funds from his department in time to satisfy the Bureau of Reclamation, then he met with the Mountain States people, and advised us to let Mountain States go ahead with their proposal and that it would be too late by the time we could get REA funds in here, perhaps six or eight months.
>
> All in all, Flaming Gorge appears to be the help we need in securing adequate telephone service to our area at a minimum of cost to our subscriber, and at no more expense to the Bureau of Reclamation than any other way. I hope that you can help us in some way to obtain what we feel is no more than rightfully ours.

The Certificate of Necessity and Convenience filed by Union Telephone as part of the REA loan application process helped Howard convince the Bureau of Reclamation that Flaming Gorge was Union's service territory. Union had an established right to serve the area, and they refused to give up that right. Mountain States eventually withdrew its proposal to the Bureau of Reclamation, who awarded Union Telephone the work for the Flaming Gorge Dam. AT&T probably had its first inkling of regret about not purchasing Union Telephone when the major company had the chance in 1948. The REA loan was approved in late 1956, and Union Telephone began its work on the Flaming Gorge Dam project.

An undertaking of such magnitude required the labor of thousands of people and the employment of a variety of contractors. Before 1957, there were but a few ranches and a small community in the vicinity of the proposed Dam site. Dutch John Flat in Daggett County, Utah, was chosen for the base of operations for construction. Trailers and facilities for roughly three hundred workers were moved in, and a US Post Office was established (see Dunham and Dunham 1977, 332–314). Just two years later, Dutch John had grown to a population of roughly 3,500, making it one of the largest towns in the area.

The terrain around Flaming Gorge is extremely rugged. The landscape is exceptionally varied, as it is crosscut by the Green River, numerous canyons of tributaries, and mountains. Drawing a beeline from Dutch John, Utah, to Mountain View, Wyoming, creates a path roughly fifty-eight miles in length. Dutch John, the starting point, is at roughly 6,400 feet above mean sea level. Mountain View, the terminal point, is at roughly 6,700 feet above mean sea level. However, the route is in no way a simple climb of three hundred feet. What follows is a very general description of how the terrain lies along a direct line from Dutch John to Mountain View.

This path has approximately 1.7 miles of vertical change in the fifty-eight miles of horizontal distance, which is roughly a foot up or down for every forty feet of travel on average. If it were possible to build a viable line absolutely straight (in a state where even the railroads change alignment to get through the rugged terrain features), a cable would be roughly sixty miles in length to account for the horizontal and vertical distances. To follow the lay of the land and install a cable that could be maintained, a ninety-mile-long cable would be needed to skirt topographical irregularities and avoid areas of potentially hazardous seasonal conditions.

Since much of this area is within the Union Pacific Checkerboard, gaining permissions to build could have triggered additional route changes. With these considerations in mind, it is easy to see how spanning fifty-eight miles in a straight line might take as much as one hundred miles of cable. This is not an unusual situation in Wyoming, as all linear structures require significant additional lengths in conforming to the realities of topography and the complex land-ownership patterns. This factor can create a formidable barrier to constructing and sustaining an efficient telecommunications network. Union Telephone planners repeatedly had to balance the physical constraints of the topography against the restrictions placed in their way by safety concerns and the requirement to obtain permission for access to build.

Area Traversed	**Altitude** (in feet above mean sea level)	**Altitude Change** (in feet)	**Cumulative Change** (in feet)
Start- Dutch John Facility	**6,400**	**0**	**0**
Dutch John Draw	**6,200**	**Down 200**	**200**
Dutch John Mountain	**6,900**	**Up 700**	**900**
Antelope Flats	**6,000**	**Down 900**	**1,800**
Green River	**5,900**	**Down 100**	**1,900**
Lucerne Valley	**6,100**	**Up 200**	**2,100**
Pico Mountain	**6,800**	**Up 700**	**2,800**
Henry's Fork River	**6,500**	**Down 300**	**3,100**
Cedar Mountain	**8,300**	**Up 1,800**	**4,900**
East Canyon	**8,100**	**Down 200**	**5,100**
Cedar Mountain	**8,300**	**Up 200**	**5,300**
Middle Canyon	**7,900**	**Down 400**	**5,700**
Cedar Mountain[6]	**8,300**	**Up 400**	**6,100**
West Canyon	**8,000**	**Down 300**	**6,400**
Cedar Mountain	**8,300**	**Up 300**	**6,700**
Dry Creek	**7,300**	**Down 1,000**	**7,700**
Sage Creek Mountain Breaks	**7,400-7,200**	**Up and Down 200**	**7,900**
Cottonwood Creek/Bench	**6,800**	**Down 400**	**8,100**
Leavitt Creek	**6,700**	**Down 100**	**8,200**
Crooked Canyon	**6,800**	**Up 100**	**8,300**
Little Dry/South Creeks	**6,600**	**Down 100**	**8,400**
Tipperary Bench	**7,000**	**Up 300**	**8,700**
End- Union Telephone Facility	**6,700**	**Down 100**	**8,800**

[6] *Cedar Mountain appears several times, as it is a mesa with fingers extending like a hand. The straight path of a line would take it across four fingers of the mountain.*

Microwave technology made sense when attempting to get a telephone signal across such diverse and rugged terrain. Microwaves are radio waves with wavelengths between one and thirty centimeters. The parabolic antennae for microwave reception have been a familiar part of the landscape since the 1950s. Using the line-of-sight or near-line-of-sight from the transmitter, this technology sends information in a tight beam for short distances, but it can be "daisy-chained" to cover longer distances by adding additional repeaters along the path between the original transmission source and the destination source. When daisy-chained, repeaters can take the signal to a distant point visible in a direct line, and from that location to the next distant, visible point extending out from the origination point, and so forth.

Even using microwave technology, it required significant effort to get around the obstacles created by the terrain. The original service line to Dutch John went from Mountain View to Hickey Mountain (14.1 miles), continued to Dutch John Mountain (44.4 miles), and from there it arrived in Dutch John (2.2 miles).

The original system configuration was changed in the 1970s. In the revised route, the installation of microwave apparatuses at the switching station in Urie, Wyoming, sent the signal to a microwave repeater on Hickey Mountain, Wyoming (17.1 miles). From there, the signal traveled on to Dutch John Mountain, Utah (42 miles), where another repeater daisy-chained the signal to the Manila, Utah, facility (15.8 miles). From Manila, a line-and-pole system took the signal to the community at Dutch John, Utah.

Building the telephone system in Dutch John was a challenge. Flaming Gorge is a major geographical barrier to get around, and the Green River had few improved crossings. Development in the area was sporadic. Once Flaming Gorge Dam was constructed, there were segments of paved road to supplement a hodgepodge of poorly maintained gravel and dirt roads. The rising reservoir pool behind the dam inundated sections of the better roads and created new disruptions in the primitive network of roads.

Graveled roads were a rarity at that time, and they were often maintained sporadically. A simple truck ride from Mountain View to Dutch John had the potential to be a rather adventurous experience, taking anywhere between two and a half and five hours to complete. In the early years of Union Telephone

Company's work in the Dutch John area, unimproved road segments offered the greatest challenges. When wet, many of the oil-shale roads became what the locals called "greasy." The reds, greens, and yellows found in the coloring of the shale were often distracting when the sun played on the wet surfaces. The high percentage of finer-sized particles—such as silts and clays—turned the dirt tracks into a slick mass when they were wet. Vehicles, even those equipped with four-wheel drive, tended to "fishtail" because of the differential weight over the wheels. The front wheels, with the weight of the engine directly over them, tended to grab the road a little better, and the rear of the vehicle spun out to the side. The sight of a truck turned sideways as it continued traveling along such roads was common if the roads were wet.

With a sufficient amount of rain or melting snow and a brief dust storm, the elements sometimes made travel a series of ambushes. Sometimes, while driving along an apparently dry road, a person's vehicle could suddenly lurch to a halt while the wheels continued to spin, which meant the driver had found a patch of road with "no bottom" to it. Usually, the only way out of this predicament was to hook a cable on to something on drier ground and pull the disabled vehicle out of the trap. On-road driving in the Rocky Mountain West often resembled off-road driving in other parts of the nation.

Snow removal was rare on most of the roads. High winds sometimes blew the roads clean of snow before the drifts had the chance to freeze into an icy mass. Once drifted with deep snows, rural roads were often not clear of snow for the entire course of a year. Snowdrifts, when in the process of melting, often created a surrounding halo of deep mud, a potential vehicle trap every bit as effective as the snowdrift itself. The combination of a bottomless road and a frozen wall was sufficient enough to deter travel using anything but the most aggressive of vehicles.

The most time-efficient manner to cover the miles between Dutch John and Mountain View was by plane. Howard became a pilot and made good use of a small airstrip built by the federal government near Dutch John. In 1963, Union Telephone purchased a plane for commuting to and from Dutch John, realizing substantial savings in labor costs as a result. Even with a plane, one still has to have Mother Nature's permission to fly in Wyoming. High winds, afternoon thermals, and storms can bounce a small plane around with sufficient violence that the passengers feel like a basketball being dribbled up the court. Additionally, landing on a muddy or snowy airstrip was not a prescription for a long and happy life.

Union Telephone prepared facilities and infrastructure to provide service to the United States Bureau of Reclamation facility at the dam, overseen by the Bureau of Reclamation's (BOR) chief in Salt Lake City. An REA loan provided the funding for the equipment used. The company also installed fire alarms at the Dutch John "man camps," the temporary housing sites for work crews. Union Telephone contracted outside sources for construction of the power and telephone lines, bringing in a Bridger Valley Electric Engineer, Amos Jackson, to help with the design. As part of the project, the company completed lines into Linwood, Utah, unaware that the reservoir was slated to inundate the entire community.

By 1958, Union Telephone had almost five hundred telephones—plus pay stations—hooked up in the Flaming Gorge/Dutch John area alone. Things were moving fast. Union was adding employees, and this growth produced a series of changes in the company. At Howard's urging, the board of directors offered Union employees health insurance, and they also instituted its first employee-relations policies. Equipment was added to the company's inventory. Howard was alert for additional opportunities, such as acquiring a half interest in the Central Utah Telephone Company.

REGULATORS 'R' US

Union Telephone benefited from the Federal Communication Commission's aid-in-construction rules. These rules are approved by the state's Public Service Commission or, depending on the state, the Public Utilities Commmision. Qualifying companies are allowed to recoup part of the cost of extending lines to customers. These rules required a construction allowance that needed to be exceeded before a company required the customer to provide an aid-in-construction fee. Basic utility laws provided that a utility with a monopoly is subject to rate regulation; however, the rates had to be sufficient to allow the utility to cover its costs, plus a reasonable rate of return on its investment. This required that the investments be used, as well as useful. This puts the company in the position of being in a "cost-plus" business. However, with the advent of competition, the arrangement could leave a utility unable to recover its investment.

Essentially, the cost-plus arrangement allowed the telephone company to recoup from the benefiting customers the high costs of extending services into the outback. As long as Union Telephone Company could justify the expenses in building the landline systems, they could pass along those costs to the customers. Union Telephone still had to pay its bills, but there were protections in place to ensure that it and similar companies were unlikely to become insolvent due to their attempts to expand services in rural areas.

Additionally, Union Telephone received a discount on the AT&T long-distance rates, based on the 1913 Kingsbury Commitment. The long-distance giant lowered its rates in order to not put the small telephone companies out of business under that decades-old obligation.

Union Telephone found itself navigating a labyrinth of rules from both the FCC and the states' public-service commissions. Also, each of the federal agencies that managed land (as well as others that did not) had their own sets of regulations and rules with which Union had to comply. Communities, counties, grazing associations, and even Union Pacific Resources—the railroad's land-management division—further added to the list of potential regulators. The process of getting approvals to build in the outback was turning from one with a few speed bumps into a high-hurdle race.

NUMBER PLEASE

The public "face" of most telephone companies came to the customers as a voice over the phone. Whether the soft voice said, "Number, please," "How can I help you?," "Operator," or just plain "Hello," the greeting started one of the most familiar customer-service rituals in history. For over a century, telephone customers expected to find the telephone operator competently serving at her post, waiting to complete the telephone connection to the right destination or to provide services for other customers. For the most part, the public tended to take their services for granted, but they were an important part of life.

The Union Telephone switchboard.

Since the turn of the twentieth century, customers came to expect a woman to answer when they rapidly tapped the receiver several times to alert the operator that help was needed. Early experiments in New York City using Bowery Boy types (street toughs) as operators were so disastrous that exchanges all over the nation sought genteel young ladies for the work. So ingrained in tradition is the role of women in this profession that Union Telephone has yet to receive an application from a male for an operator position. That is not to say that women have only operated the switchboards. It took decades for the event to happen, but eventually some Woody men, like generation-three member Lee, took a turn at running the board. It may be that some customers were surprised to hear his deep voice responding to a call for directory assistance.

Early in the company's history, John D. established a building to house the operators, who were expected to stay on the premises overnight to fulfill their long shifts. The Woody women served more than their fair share at minding the switchboard throughout the decades. When several members of the Woody family left for Sweet Home, Oregon, to care for Isabell during her tuberculosis treatments, Millie Davis and her husband, Jeff, occupied the operators' building.

The operators provided both customer and public services. Typically, operators were the first leg in the primitive emergency-response systems of rural Wyoming, Utah, and Colorado. Lives and property were saved because the operators were there to take the call and pass them on for assistance to the proper destinations. The modern 9-1-1 emergency-response system was years off in the future, and, absent a unified system, telephone operators worked out individual protocols with local and county emergency teams. This was an important part of the job.

Operators also helped with the collateral emergencies created by an incident. For example, when the ambulance was dispatched to the aid of a heart attack victim, the operators at Union Telephone spent hours finding relatives, arranging for the feeding of the victim's animals, and even finding a safe place for the children to wait while the adults were all in the hospital dealing with the medical crisis. It was a fine example of how neighbors take care of one another during emergencies, and Union prided itself on being a good neighbor.

Operators dealt with the lonely, the elderly, the infirm, the lost, and the inebriated. They were asked for recipes, directions, dates, and recommendations for food and lodging. They suffered more than their fair share of sexual—and other forms of—harassment. They were asked to remind customers of appointments and to take their medications at the proper time. Sometimes, they were tasked with seeing a customer through a rough day and reminding troubled individuals that life indeed was still worth living. Many services that the operators performed as "unofficial duties" of the job were later taken over by various social and charitable services, as well as electronic personal assistants. The Hello Girls were trained to handle all of these situations in a professional, efficient manner.

If someone called using a Union Telephone payphone to reach another party, the operators made an effort to find that person and get them to the telephone to take the call. That might be convenient and possible if the person being

MS. INFORMATION

The following call coincided with a QWEST system fiber-optics outage.

An elderly lady called in and said: "Honey, I can't call anywhere in the US. Everything I call says it's been disconnected."

The operator explained this was just a result of the fiber-optic lines being inoperative.

The relieved caller replied: "Thank God. Now I have to go back and erase all the scribbles in my address book. I thought everybody had moved."

sought was at the local bank, the garage, the market, or the bar down the street. However, not every call was simply completed, and the operators literally went that extra mile in the name of providing excellent service for their neighbors.

The operators apparently had a lot of fun when regular customer and Union board member George Smith used one of the booths. George, a local rancher, always seemed to be puffing away on a big cigar while in the booth, filling the small space with a dense cloud of pungent white smoke. The operators jokingly placed bets on whether or not George was likely to suffocate himself in the thick billows of smoke within the tiny booth. He never did.

The telephone office was across the alley from a bar, a circumstance that created both humorous and tense moments for the operators. Sometimes, customers who were "in their cups" needed assistance with calling home for a ride. In other instances, drunkards came to the Union Telephone Company office to seek a refuge from the cold. The operators found themselves with the unpleasant duty of having to throw the noisy, unwanted guests out of the office. Predictably, some of them staggered back into the office shortly thereafter and had to be ejected more than once.

In an attempt to remedy this, Union Telephone placed a fence in the alley to discourage the drunks from disrupting the office. This plan backfired, and the fence became a favorite place for the drunks to relieve themselves. The operators often chuckled as they heard the familiar sounds of this unapproved and unwanted use of the company's fence. Ever the trickster, Bonnielee convinced Esther to heave a bucket of dishwater over the fence to discourage the continuance of this behavior. It is not known if the sudden downpour of soapy water had the desired effect, for Bonnielee and Esther were too busy laughing.

In the 1950s, Union Telephone was having difficulty finding sufficient operators in the Mountain View area to fill all the shifts. Union installed a dial-up switchboard in 1956, effectively eliminating the need for customers to complete a call through an operator. At this time, Union reluctantly contracted both the long-distance operator and the information functions to its rival, AT&T. Bell-system operators replaced the Union Telephone workers. Looking back on this,

Howard explained: "The operators are the only touch you have with the public. It was a mistake that they went away." It took roughly thirty years for the mistake to be corrected.

In 1957, the last of the old magneto phones were replaced by dial systems. Instead of having to connect through the operator, customers dialed directly to another customer. Dial phones were not new technology at the time. Commercial models of dial phones were available as early as 1905, and AT&T had installed its dial system during the late 1920s. Bell Telephone had introduced the now-familiar "French" telephone design in the late 1920s. The cumbersome units with the speaker/dialer and receiver were replaced with a single handset that was separate from the dialing apparatus. Replacing all of the telephones was an expensive proposition for Union Telephone, but the time had come to complete the upgrade, which was made possible by the availability of the REA loan.

The telephone company owned all of the bulky, metal dial telephones. On these older-model phones, the dials tended to not rotate easily. An unintended result of this feature was that the dial phones were somewhat child resistant, as a toddler playing with the phone might have to use both hands to rotate the dial. Because of the difficulty in rotating the dials, customers often kept a pencil near their telephones to assist with the dialing process. Dialing was, at best, a slow process, only made more palatable by having to dial five numbers for a local exchange. If the caller needed help with making a call, dialing "0" connected one with the familiar operator.

Union installed new switches to convert their system to a dialing system. The Union Telephone Company established its dial capabilities prior to the service provider in Evanston, as well as many of the other exchanges in the region. As it often happened, Union Telephone's consumers requested specific services and upgrades. In one instance, individuals from Christmas Meadows, Utah, asked if the company could extend dial service into their area. The Christmas Meadows residents were informed that they needed to contact their own provider (the one located in Evanston) and ask for the service.

Union Telephone eventually served the community at Christmas Meadows. The prodding from that community's customers apparently spurred the

Evanston telephone provider to switch to dial service following Union Telephone's lead. Union Telephone provided service to Christmas Meadows by employing a line purchased from Mountain Fuel. During the centennial year for Union Telephone, Christmas Meadows landline service was served by a microwave system provided by Union Telephone.

BURIED TREASURE

Howard operating the heavy equipment used to plough telephone cable into the ground.

According to Howard, a major accomplishment during his tenure was to take all of the open lines and bury them. These buried lines are surrounded by heavy insulation. In the old days, a trencher or backhoe opened a small trench into which the shielded line was placed and buried. As technology improved, it was possible to plow in the line with a machine that excavated the trench and simultaneously fed the line into the trench from a reel. Modern equipment buries the line as it is laid. Burying the lines significantly reduced the need for repairs due to storm damage. While the lines are always subject to being cut by careless excavation, buried lines create a far more stable telephone system. The plowing in of the Union landlines began in 1972 and finished in the 1980s. However, as Union subsequently acquired new exchanges, the process continued into the twenty-first century. By the time of Union's centennial celebration, all of their landlines were underground.

REACH OUT AND TOUCH SOMEONE: SURVIVING THE SIXTIES

The 1960s were a period that witnessed of some of the most rapid cultural and economic changes in the twentieth century. The United States was deeply involved in a cold war against the Warsaw Pact nations, which was led by the Soviet Union and the breakaway power of the People's Republic of China.

The threat of imminent nuclear oblivion loomed large in the public consciousness, and a crisis in Cuba almost tipped the world into a third world war. The war on poverty was put on hold to fund a shooting war (another in a series of "police actions") in Vietnam. Race riots, war protests, strikes, terrorism, assassinations, and drug experimentation filled the headlines. The first moon landing, technological growth, increased tolerance, voting rights, artistic achievements, and the last large-scale expansion of both America's infrastructure and the manufacturing segment were among the positive chapters of this period.

At the time, Wyoming did not fully share in the benefits of the nationwide economic boom of the 1960s, an event fueled by military and space-program spending. Ranching and tourism were still important, but even the modest level of increased energy development had outstripped these more traditional legs of the state's economy. Wyoming was in the familiar position of waiting for something to happen.

An economic mainstay of development in southwestern Wyoming was the growth of the trona (soda ash) industry. Trona, a naturally occurring form of sodium sesquicarbonate, is used in the pharmaceutical, glass making, detergent, gas desulphurization, and baking industries. Baking soda is a form of pure soda ash with which most people are familiar. The lands between Granger and Green River, Wyoming, contained roughly ninety percent of North America's known reserves of trona, which was first discovered in the region in 1937. By 1947, Westvaco Chemical Corporation was commercially producing the compound from a 1,500-foot-deep mine.

During the 1960s, there was a marked increase in trona production. New and expanded mines meant railroad facilities, trucking facilities, corporate offices, processing plants, construction facilities, and expanding local communities. Union Telephone was perfectly positioned to serve the expanding needs of this industry, along with those of the related businesses that supported the mines. Much of Union's focus during this period was aimed at the growing needs of the international corporations that were entering the area as part of this soda ash bonanza. The trona expansion had longevity to it, and it continued through the 1970s. In Union Telephone's centennial year, trona mining remained an important leg of the region's economy.

National telephone systems were in a state of flux during the 1960s. It was a time when the old alphanumeric system, long used for making calls, was changed to an all-number calling system. Prior to the changes, if one wished to reach a certain person, the caller could dial or ask the operator for a telephone-exchange name followed by a four-digit number, such as "Tremont-3248". The all-number calling system changed this to 875-3248. For interstate calls, a three-digit prefix was added.

The times were changing the nature of the telephone into telecommunications. As the 1960s was the decade of space exploration, it is little surprise that the communications industry was entering space. The first communications satellite, Telstar, was an active system much like a microwave tower in space. Signals sent to the orbiting repeater were rebroadcast back to earth. This communications satellite paved the way for others to follow, and these satellites have become an integral part of twenty-first century communications. During the cold-war politics of the 1960s, Congress passed the Communications Satellite Act, which established a quasi-governmental corporation, Comsat, to oversee how the communications companies used space. The opening of space to commercial telecommunications had far-reaching effects for the industry, which in Union's centennial year had become of prime importance in the globalization of communications.

Union Telephone continued its growth and modernization during the 1960s. The upgrading of services was continuous throughout this period. Another highlight was the transition to the new dial-tone system. The company had already built an exchange in Dutch John with microwave capability to assist in the continued servicing of Flaming Gorge. Union Telephone crews stayed in the new exchange building during its construction because there were no local accommodations.

TAKING STOCK

I liked them, but didn't like their way of doing business.
–Howard Woody

As the company became profitable, it grew increasingly difficult to buy back outstanding stock. Howard and John D. discovered that they were not the only parties intent on acquiring the remaining shares in the company. In the early 1960s, Howard learned from a stockholder that the secretary of the board and the company's attorney were attempting to buy up shares of stock in order to acquire a controlling interest in the company for themselves. According to Howard: "[The two of them] devised a plan to get me fired. I knew that."

The schemers completely underestimated Howard, who calmly and quietly collected large numbers of proxies for the next annual meeting of the stockholders. When the time came to register the proxies, Howard presented stack after stack of signed proxies for the secretary to register. Howard took satisfaction in observing the faces of his opponents when the realization hit them that their plan had failed.

Following that board meeting, the two presented Howard with an offer to sell him their shares in the company at an inflated price. Howard wanted to get their stock, even at such a high price, and he agreed to the deal. The three men subsequently met to sign a contract for the transaction. Clearly not giving up on their desire to take over the company, the two proposed an exception-

MS. INFORMATION

Operators get some of the strangest calls. The following is an example of an actual call to the Union switchboard.

A caller asked for a listing and the operator replied, "Would you like to be connected or just take down the number?"

After a long pause the caller replied, "Well, let's see . . . I'm not sure what I should do."

ally complex contract whereby Howard had to place his shares in escrow and personally make monthly payments at the Evanston Stockmen's Bank on certain dates, between the hours of 10:00 a.m. and 4:00 p.m. The contract did not allow Howard to pay the loan off early.

Any breach of the payment schedule would result in Howard losing the shares in escrow. The contract apparently was designed knowing that in Wyoming's difficult climate, Howard might have significant issues getting to Evanston at the prescribed times and would hence would lose his shares. During the winter months, the roads between Evanston and Mountain View were often closed due to deep snows. Unable to accomplish their goal with a vote by the stockholders, they appeared willing to rely upon Mother Nature to hand control of Union Telephone to them.

Once again, the two men had seriously underestimated Howard. As a member of the board of the Uinta County Bank, Howard knew the Evanston Stockmen's bank president. The two worked out a method to assure that payments were submitted on time and according to the odd terms of the contract. Howard made the payments, as required under the contract, for the first few months. He then approached the two associates in the stock deal and offered them a lump sum payment. Howard believed a single payment would be hard for them to resist. The two did not agree at first, but eventually the lawyer convinced the secretary of the board of directors to take the deal. It was pricey, but Howard had the stock clear of encumbrances, and with the stock came clear control of the company.

The secretary of the board apparently was not content to take his profit from the sale and let matters lie. The unhappy schemer attempted to discredit Howard with the Rural Telephone Association by allegedly submitting a bogus tip to the federal regulators. This resulted in an audit of the company's finances. Nothing came of this muckraking. When the audit was completed, Howard recalled the auditor telling him: "I have looked at every transaction, action, practice, and document that this company has made since it was incorporated. I can't find anything to suggest any wrongdoing." Once again, the plans to get rid of Howard were foiled.

With Howard holding sufficient shares and proxy votes to control the company, it was somewhat of a surprise that there was continuity on the board. A lesser man might have used that power to throw the secretary off the board of directors and put in a more loyal person. Howard kept the disloyal officer on despite the recent history between them. According to Howard: "He has learned a lot about the telephone business. It would be costly to replace him. I didn't know anyone that would be better. Besides, I believe an old dog knows not to suck eggs."[7]

Difficulties with the board's secretary continued over the years. Howard joked about how the secretary was most excited when there was a proposal for Union to purchase its own airplane, because the secretary was very much in favor of Union purchasing a plane. The purchase originally was proposed so that Howard could cut travel time to places such as Dutch John. However, there seemed to be more to the secretary's support than a desire to make more effective use of Howard's time. As Howard jokingly recalled it, "He was hoping that I would crash." When the secretary continued to try and undermine Union Telephone's relationship with REA, Howard's patience finally came to an end and the secretary of the board was removed after roughly three decades with the company.

ASK AND IT SHALL BE GIVEN

A recurring theme in the Union Telephone story has been the lengths to which the company went to meet the needs of its customers. Time and time again, customers came forward and asked Union Telephone to provide specific services. And time and time again, Union Telephone listened, researched the possibilities, and provided the requested services when feasible.

The Wyoming economy was experiencing an expansion of oil and gas production. Howard recalled that oil-field manager D. C. Wilson approached him for phone service in the Sage Creek Mountain area. As the gas and oil industries were developing in areas like the Moxa Arch, near Granger, Wyoming, the

[7] *This is a play on the old idioms warning people not to "teach their grandmother to suck eggs" and that "you can't teach an old dog new tricks." They mean very much the same, as grandmother already knows how to suck eggs and old dogs don't learn well.*

energy companies expressed a strong desire to get service to their facilities in those remote places. As Howard recalled: "Someone asked for service and we had to find a way to provide the service. We searched and we searched and we searched until we had something to offer."

Considering Mr. Wilson's request, Howard remembered reading about radio-telephones being offered by Motorola. He had heard that logging companies were having success with the use of radiotelephones in Oregon. Howard contacted Motorola and traveled to Oregon to witness the systems in operation. Based upon this visit, he decided the radiotelephones would work in the Wyoming gas and oil fields.

Radiotelephones, also known as Improved Mobile Telephone Service (IMTS), turned out to be a revenue boom for Union Telephone. The company first installed its IMTS in November of 1962. As Howard Woody recollected: "Communications was so vital to them, boy, I'll tell you. They spent money like you can't believe. That was the first thousand dollar telephone bill I've ever seen." Union Telephone continued a long and healthy association with the energy industry, and they were quick to establish service as gas fields sprang up in the 1980s and the successive decades.

Motorola manufactured radiotelephone systems as early as 1958, beginning with its Motrac radio designed for vehicles. In partnership with Illinois Bell Telephone, they introduced their pioneering mobile telephone system on October 2, 1946. The radiotelephones of the period were extremely large and bulky. As battery-operated units in the years when battery technology relied heavily on lead, each radiotelephone essentially was a satchel that weighed several pounds. Because of the difficulty in carrying such units, it was common to have the cumbersome equipment bolted into the cab of a vehicle, where it was then powered by the vehicle's battery and engine. These burdensome giant communication devices paved the way for the smaller radiotelephones that became cell phones.

Union Telephone purchased an IMTS system from Motorola and placed a microwave tower on Hickey Mountain, which carried a signal to Union Telephone's first radio tower. The signal from the radio tower was sent back to Urie, where the signal was then available worldwide. This tower was followed some time later by Union's installation of another repeater on the Bear River Divide between Kemmerer and Evanston. The expansion into radiotelephones created opportunities for the company, and it allowed the corporation to provide service to remote, special-needs customers.

There are few industries that move as fast and furious as energy companies during the middle of a gas boom. Radiotelephones provided the flexible communications necessary for the gas industry to construct and maintain the frantic growth of gas facilities in rural Wyoming. As the energy industries expanded, Union Telephone was well positioned to expand the radiotelephone coverage in Kemmerer, Granger, and La Barge. Union provided IMTS coverage focusing on the southern fields, only providing occasional services to the more northerly reaches of what the locals have named "gas patch."

As drilling rigs flooded into southwestern Wyoming to drill hundreds—and eventually thousands—of new wells each year, Union Telephone scrambled to keep up with the industry's needs for reliable communications in remote locations. As long as there was a line-of-sight between the worksite with a radiotelephone and a phone tower, there was a link to the broader telephone system.

Technology in the 1960s was plagued by the need for high levels of maintenance. Before transistorized radio sets, the vacuum tubes which made radios work were fragile and required frequent replacement. In the late 1940s, a team of Bell Laboratories scientists invented the transistor. The earliest transistors were fidgety, but by 1955 the technology had sufficiently improved. While the technology improved rapidly, there was a lag between invention and marketing. Union employee Jim Newland worked in the field almost exclusively replacing the old vacuum tubes in the fragile radiotelephone units into the 1960s.

Radiotelephones, like most others in service at that time, were not for everyone. IMTS use was cost prohibitive for what was, in essence, a large party line. What was needed was a way to make radiotelephonic communication more similar to landlines.

CONNECTING THE FIRST LADY

As a result of the company's progressive installation of radiotelephone services into the Dutch John area, another opportunity presented itself in 1964. That year, the First Lady, Lady Bird Johnson, officially opened Flaming Gorge Dam. Because it had operating radiotelephone service in the Flaming Gorge area, Union Telephone Company was asked to provide communications for the First Lady and her entourage. Howard installed a radiotelephone system in one of the motorcade's official vehicles, which was used throughout Mrs. Johnson's visit. Howard followed the motorcade in his truck with another radiotelephone for as long as the First Lady was present. He remained available to ensure continuous operations of these delicate, high-maintenance radio systems. When the First Lady's motorcade finished its visit, Howard was on hand to remove the radio-telephone, thus ending one of his most memorable customer-service calls.

THE POWER OF NEIGHBORS

Connecting communications facilities to power systems sometimes was an issue for Union. In an attempt to provide television service to Bridger Valley customers, Howard approached the railroad about getting power to a translator on Medicine Butte that could broadcast television into the Bridger Valley. The Union Pacific Railroad had a repeater site on Medicine Butte about a quarter of a mile from Union's proposed translator site. It made sense to come off the existing line on the butte rather than build a whole new system. Howard asked for permission to tie into Union Pacific Railroad's power line.

Throughout the decades, Howard and Esther ran the company hand in hand.

Howard picked up the story from here. "As it turned out later on, the company was trying to get television service on Medicine Butte and couldn't get power up there because it was way too expensive to build power lines up there. The railroad had power lines up there, but the man in charge of those lines told me

I couldn't use them. I called Bob Brown and said, 'Bob, do you remember Perkins said that if I ever needed help to let him know? Well, I'm going to have to call you on it.'"

Brown asked what Howard needed, and then replied, "Give me three days, and then go back and ask them again." Howard recalled: "Well, I waited three days and asked again. They said I could hook up to that line like nothing else had ever been said! That was the last dealing I had with Bob. The next thing I heard, he was dead."

Calling in these neighborly agreements to help one another was often not easy to do. In true Wyoming fashion, the pledge was honored, and shortly thereafter Union Telephone connected the translator to the grid. Months went by after the connection was made and Howard had not seen a bill. When he called Bob to inquire what Union Telephone owed, Bob informed him that his good neighbors at the Union Pacific Railroad provided the power *gratis*. Not until Utah Light and Power took over the service to the area from Union Pacific Railroad did Union Telephone receive a bill for power to the translator.

THE CALM BEFORE THE STORM

The Union Telephone system expanded during the 1960s, and soon it was extending lines to more Colorado and Utah communities such as Lodore, Meeks Cabin Dam, Christmas Meadows, and Brown's Park. A central office building was constructed in Greendale. The company had been upgrading by making a major purchase of equipment (including a Motorola microwave system) from Mountain States Telephone and Telegraph.

Twenty residents of Brown's Park had approached Howard and asked for Union to provide service to their community. This contact resulted in the 1968 Union extension of service into Brown's Park, which was among the most colorful projects undertaken to date by the company. The area was rife with local color dating back to Tom Horn and Butch Cassidy, but the attraction for Howard was the chance to provide services to his neighbors who needed communications capabilities.

So remote and rugged was the route to Brown's Park—even by the local standards of the frontier—that this project was one of the more demanding construction projects for the company. The surrounding mountains blocked access to Brown's Park Valley, forcing service to be routed through Red Canyon. The work was done by two of Union's contractors from Uinta Basin, who constructed an open-wire line that expanded the number of lines available within the service area and connected many of the local ranches. Union Telephone crews replaced this open-wire line with an underground cable during the 1980s.

THE 1970s ENERGY BOOM

On October 6, 1973, Syria and Egypt attacked Israel and set off what became known as the Yom Kippur war. When the United States provided airlifted supplies to the beleaguered Israeli Defense Force, the Organization of the Petroleum Exporting Countries (OPEC) announced a series of cuts in oil production, combined with price increases. The United States, Western Europe, and Japan were all declared hostile countries by OPEC, and for these nations a total boycott was declared. The United States, deep in a stock market crash since early in 1973, was ill prepared for what followed.

The OPEC embargo had far reaching effects for Wyoming and Union Telephone. It was one of the prime movers in setting off an oil and gas boom. Most of the boom's activity focused on known areas of production like Wyoming's Overthrust Belt and the Anticline. Places like Granger, La Barge, Green River, Rock Springs, Rawlins, and Evanston found that they were unable to provide goods, services, and infrastructure sufficient to support a massive and seemingly overnight increase in the population. Counties like Uinta County, located in the heart of Union Telephone Company's service area, almost doubled in population over the span of a decade.

New workers flooded into a region that only a few years prior had qualified as a frontier. For Wyoming, it was the beginning of the end of its frontier days. Tiny communities scrambled to increase the size of municipal services such as water, electricity, medical facilities, law enforcement, sanitation, schools, and communications. Even with the importation of a massive number of manufactured homes, the housing market could not handle the influx. Where once there was open sagebrush, massive trailer courts sprang up. Entrepreneurs took advantage of the housing shortage by renting out back yards, parking lots, and alleys to the migrants who were desperate for a place to stay. Even with such creative measures to ease the shortages, "man camps" sprung up in remote areas and near construction sites. Many families set up tents, built crude shacks, and even slept in caves in the Wyoming backcountry. These "squatter camps" recalled images of the temporary housing of emigrants fleeing the dust bowl and the Hoovervilles of the Great Depression.

As a utility provider, the gas boom of the 1970s was a major growth opportunity for Union Telephone Company, which was positioned right in the middle of the economic explosion. Well-financed companies, speculators, and well-paid workers flooded Union Telephone's service area. Service industries sprung up to meet the needs of this burgeoning population. As workers flocked to higher-paying jobs in the gas fields, there was a local labor shortage.

Howard operating a backhoe/front-end loader.

During the boom years, demand for telephone service skyrocketed. Union Telephone Company was hard-pressed to install sufficient lines and hook up new subscribers. The demand for services exceeded the company's ability to meet it. Not until 1990 were there more than twenty-five company employees at any one time to handle the work. The handful of Union Telephone employees available in the 1970s struggled to install new lines in places already within the service area, maintain existing services, and expand service to newer, more remote locations. It is during the boom where generation three made its first substantive contributions to the Union Telephone story. Their energy and work ethic were greatly needed.

The federal government encouraged additional domestic oil and gas development as dwindling energy supplies became a strategic issue. However, Wyoming was not the kind of place that adjusted easily to unbridled industrial expansion. The climate itself served as a speed bump, slowing development by restricting the months available for construction. Several other factors restricted development. Primary among these is the limited availability of water. Water was necessary for construction, road maintenance, drilling, fracking, and for the survival and health of work crews stationed at man camps. Established towns, such as Granger and Rock Springs, had long histories of intermittent water shortages caused by the unreliability of flowing water in the natural drainage systems. In such an environment, the man camps and places where development was occurring often had to have water hauled in at great expense, and they were on their own hook to develop manners to deal with wastewater. In an area with primitive road systems, the outlay for hauling water could be cost prohibitive.

Even in the early twenty-first century, there were far more miles of dirt roads than paved roads in Wyoming, a situation that made supplying workers and industries a major concern. For much of the Cowboy State, snow potentially occurred in almost every month, which made the crude road system tenuous at best. Industries were forced to invest heavily to establish reliable roads into remote areas if they were to develop those places. Once the roads were completed, keeping them open and usable became the next challenge. Municipal, county, and state budgets were already stretched thin, and they were insufficient to build and maintain adequate roads to sustain the level of activity created by the boom.

During the 1970s, Howard continued to serve on various boards of the developing telecommunications industry. Howard (seated, center) served as an officer for the Rocky Mountain Telephone Association.

Wyoming's complex pattern of land ownership became a factor in keeping the 1970s boom from being more permanent. Access to work sites and suitable logistical areas was often difficult and expensive to acquire, and federal lands had different restrictions and rules than state or private lands. The vast lands held by Union Pacific Railroad (which were managed by their resources division) had its own bureaucracy, use-rate charges, royalties, and environmental rules to deal with. Union Telephone had not encountered such issues in its pre-cellular years. However, pipeline, seismic exploration, mining, and drilling companies apparently had different experiences.

Acquiring rights-of-way in the Cowboy State was not always easy. Absentee owners were often difficult to track down. The best access route into a prospect might be in the hands of a competitor, and a less suitable route might result in additional expense and maintenance issues. A company might sabotage or inconvenience its competition by restricting admittance to mineral leases by leasing or buying critical access points to the backcountry in a very expensive game of mahjongg.

Booms tended to sow the seeds of their own undoing. As thousands of wells were brought in and product filled the rapidly expanding pipeline systems, the very abundance of the new production affected the market. Lower prices resulted from an overabundance of available product. When market prices fell below certain levels, the developers closed in wells and waited for the return of higher prices. In areas where production costs were higher than returns, systems might be abandoned or sold to companies that thought they could do it smarter and better than those who developed the same prospect in the past. When the gas boom finally sputtered out, the employers and workers moved to the next boom area. The boom/bust cycle almost seemed to have a natural rhythm to it. From Union's perspective, the boom brought high demand, and the bust reduced the need for telecommunications services.

TECHNOLOGY NEVER SLEEPS

Part of normal operations in the communications business is the constant upgrading of technology. Union Telephone Company's early microwave sites and radiotelephone sites had components that were operated by vacuum tubes. Eventually, transistors replaced the fragile tubes, but even those were soon replaced with newer, more efficient integrated circuit boards and chips. Union Telephone quickly adapted new technologies as they became available. The rapid developments in telecommunications and electronics kept the company's engineers and technicians busy as they modified the system to reflect changing industrial standards. Examples of the rapid upgrading and improving of the system included burying the open line in the Bridger Valley (1972), microwave upgrades to include analog microwave and passive repeaters (1972), and one of the first computerized billing systems (1976). The latter was made possible through computer software developed in house by Union Telephone staff, including members of generation three.

During the mid-1970s, Union Telephone borrowed over three million dollars from the Rural Telephone Bank (RTB). The RTB was established in 1971 under the Rural Electrification Act as an agent for providing loans for the development of telecommunications in rural areas. (The bank was dissolved in 2005.) The money made available to Union Telephone helped finance Union's aggressive expansion and modernization during this period, such as the construction of a new central office building in Lyman.

PUT ME IN COACH

Howard looked back on his days of running the company with pride. He transformed the company from a tiny local exchange (that had not turned a profit for decades since being founded) into a regional telephone system encompassing a service area of millions of acres. Howard's acumen and innovation ensured that company founder John D. Woody lived long enough to see the company attain profitability. It was no small thing that John D. got to see his dream realized by the continued efforts of his descendants. Union Telephone Company had attained profitability by the time generation three was ready to assume leadership of the company.

Union Telephone began operation with outdated—and sometimes homemade—technology. By the 1970s, however, Union Telephone was both including and pushing state-of-the-art technology for their system. It became a mantra for the company that people in rural communities deserved technology as advanced as the cities. It was time to see where generation three was going to take the company. Members of generation three quickly learned the ropes and were positioned to take over the day-to-day operations of the company under Howard's continued guidance.

"I never wanted to be a millionaire. I never wanted anything but to build a telephone company. I saw my dad struggle with it," said Howard.

Our next connection brings us to the fulfillment of a concept that occurred to Howard while he was working atop Hickey Mountain. It was there that Howard looked into the future of communications, envisioning a path that led to cellular phones.

Howard Woody, a man with a vision.

Chapter 4

The Third Generation: Leaps of Faith

If we just had a phone we could carry in our shirt pocket, like a pack of cigarettes, that we could talk on. That would be the ultimate. That would be the most fantastic thing that ever happened.

–Howard Woody, talking about the inspiration he received while on Hickey Mountain

As the third generation matured, they presented Union Telephone with its first real challenge in terms of the succession of company leadership. The first generation in the company had but a single member, John D. Woody, and he was far more interested in building the company than running the board that managed Union's business aspects. A tactic John D. used to enlarge the company involved giving away or trading Union Telephone stock. It was not until late in his life that there was a coordinated effort to return the shares to family ownership again. By the time generation three was ready to run Union Telephone Company, the family had acquired a controlling share of the stock. Union Telephone truly had become a family-owned and family-operated business.

Generation two had but a single male member, Howard D. Woody, and he was considered to be in a position to inherit the mantle of corporate management. In a time when women were typically relegated to secondary roles in corporate management, Howard's sisters and his wife had very important roles to play in operating the business, but they bumped into a glass ceiling regarding participation on the board. Generation three shattered this barrier.

The men of generations two and three. From left to right: John G., Howard, Jim, and Lee.

Howard Woody built upon the foundation established by his father and moved the company into profitability, expanding its overall service area. Even in retirement, Howard clearly had a strong hand in running the company, and his influence was present in just about everything the company accomplished. By having several candidates qualified for leadership positions, generation three placed the company and the family in the enviable, but difficult, situation of determining which among the children of Esther and Howard were going to run the corporation. This was not unusual for a family-owned business. How businesses dealt with succession issues often determined whether a company continued as family-run businesses or was either turned over to outsiders, sold outright, or closed.

The members of the third generation included John G. (born 1944), Gloria D. (known as Dee, born 1947), James H. (known as Jim, born 1949), Lee J., (born 1952), and Bonnie Jean (born 1958). Esther and Howard instilled a solid work ethic within all of them. Esther had the primary duty of raising the children. Like most mothers, she had ambitions for her children, and she was a driving force in fostering each child's desire to succeed. The third-generation members learned life lessons on the ranch, but none of them were encouraged to become the next Howard or Esther as much as they were guided to find their own ways to happiness in life.

Howard and Esther discouraged their children from developing a sense of entitlement. If the children wanted to take a leadership role in the company, there were unofficial rites of passage that they had to undertake. The requirements for leadership were only formalized years later, but they were eminently practical. Each potential family board member had to work their way up the corporate ladder, spending time in each of the different departments of the company. Anyone wishing to become a leader within Union's corporate structure needed to have a résumé that included a period of work outside of Union, where the candidates were expected to develop skills helpful to the family business. This preparation brought in ideas from other companies and broadened the skill sets and experiences of Union's leadership.

Howard was proud of his family. He readily bragged about the way the children accepted the torch of company leadership that had been passed from John D. to him. According to Howard, all but one of the children learned to put the business first and the family second. "I had one daughter that didn't accept that. And I said, 'You didn't think enough about it.'" Howard and his daughter did not seem to reconcile their differences on this matter, but they have made their peace.

Howard never earned a college degree. However, he was determined that the children were encouraged to acquire skill sets that might help the company. His children all went to colleges of one form or another. John G. earned a master's degree in electrical engineering. Jim earned his degree in accounting and business management. Dee studied electronics in correspondence courses, earning an associate's degree. Lee took college courses but never finished a degree. Bonnie Jean studied accounting.

Whether it was John D. Woody finding out about building a telephone system through a correspondence course, or Howard Woody learning from a magazine how to salvage a wrecked plane and build a snow mobile, the generation-three Woody family members were active in acquiring knowledge about matters that were a benefit to the company. The family members were open to working outside of their comfort zones in order to gain skills and abilities the company needed. Generation three applied a hands-on approach to the work, becoming familiar with how to do almost everything they expected of their employees. This trait has served Union Telephone Company well, and it has given management pragmatism and an ability to troubleshoot problems that getting a master's degree in business administration alone does not usually provide.

This is one of the planes that Jim Woody built.

Under a traditional succession plan, John G., as the eldest son, was most likely to take over the company. He officially started with the company after high school. John G. had been around the business all of his life, even being taken out on line-repair jobs while still an infant by his mother and grandfather. He held numerous positions in the company, starting out as a lineman and working in construction and maintenance. He then served as chief engineer, and eventually he sat on the board of directors and became the chief executive officer.

Jim served as general manager in the early 1990s. He and his wife Pamela have worked for the company for decades. Jim did much of the computer programming for the firm, playing a major role in developing software for computerized billing. He mastered the complexities of building and maintaining complex electronic equipment. In his leisure time, he built airplanes. He was also a deacon in his church. He loved to fly, and he used air travel as much as possible to commute throughout the American Outback, rather than drive through its complex terrain.

Dee Woody enjoying Union Telephone's first company picnic.

Dee was born one day after St. Valentine's Day. She was much like her mother, being roughly five foot two inches tall. Dee was the first woman to serve on the board of directors, something she was exceptionally proud of. She had a string of unpleasant marriages, but through it all she maintained her professional attitude and loyalty to the company. Dee spent several years living in Nevada and California, but she eventually returned to Mountain View and the company. She was a real scrapper who liked to compete with the boys of her generation while she served on the board. To alleviate the tensions of life, she became an avid attendee at Fort Bridger's number one event, the annual mountain-man rendezvous, affectionately called the "Vous" by the locals.

Lee was a quiet man. He was unassuming and steady. His shyness might be partly attributed to his poor eyesight. Lee was reliable and worked best in the office situations. As a board member, Lee often found himself caught between powerful personalities, and he tended to show indecision in such instances. When his sister, Dee, suddenly passed away, the dynamics of the board changed, and Lee bloomed into a real leader. He was a good listener and was levelheaded and kind-hearted. The support of the employees helped him emerge from the corporate shadows to become a force on the board. Lee was suddenly struck down as well, right when he was flourishing in his personal growth. The employees liked him as a person and a supervisor, and they fondly remembered him in Union's centennial year.

Bonnie Jean was the self-described black sheep of the generation. As the youngest, there was an age difference of fourteen years between her and brother John G. She claimed to never have attained the respect held by her older siblings. Bonnie was outspoken and opinionated, and seemed to take on the "punk kid" role. Out of all of Esther and Howard's children, Bonnie Jean sounded like the one who inherited the most of Esther's forthrightness. Very much a family person, Bonnie Jean put the family first. This was consistent with her anti-authoritarian "punk" role, but it set her against other members of the family who followed Howard's philosophy of putting the company first.

Bonnie applied for admission to the management team, but she was rebuffed. She speculated that this was because family members were concerned as to how her wildness might affect the governing of the company. Part of the issue may have been timing, as the management-team approach that was adopted in 1995 had turned the company into a meritocracy. Management-team members were expected to earn the positions, something Bonnie had not had an opportunity to do as much as her elder brothers and sister. Bonnie undertook the training to become part of the management team, but she entered into this career path so late in life that she was enrolled with the members of the fourth generation. At the time, this did not sit well with her.

Bonnie was reluctant to appeal directly to her father, so she continued to try and convince her siblings that she was ready. After years of struggling to break into the management team, Bonnie and her husband, who also worked for the company, moved out of the state shortly after Dee terminated the employment of Bonnie's husband at Union. Bonnie Jean has made her peace with what she described as her exclusion from the family management team, and has thrived in her new life. In this manner, things worked out for the best.

The members of generation three began serious work with the company starting in the late 1960s and early 1970s. Because of this, the accomplishments during the boom of the 1970s were every bit as much a generation-three event as it was for generation two. In 1971, John G. was appointed as the systems engineer and Jim was appointed as the accountant (they had worked for the company in other positions prior to these appointments). Prior to assuming full-time positions with the company, some of the generation-three members remember being "paid" to wash phones, treat poles, or other kinds of tasks at the family business when they were children. Each generation-three member went into the business in his or her own way, and for their own reasons. However, the image of their own father carrying on the work of their grandfather was powerful and clearly had a profound impact on them all.

Generation three was the first generation where working for the company and making a decent salary was a genuine option. The company finally attained profitability just a few years before members of generation three were old enough to enter the Union Telephone corporate structure.

On John D. Woody's watch, the ranch provided most of the food for the family and a small income, but the family could not break out of their poverty. By the time Howard returned from the army, the ranch had lost all its cattle and was down to a few pigs, goats, and chickens.

It took Howard decades to rebuild the cattle herds. Howard clearly loved running the ranch, and he has managed its operation since 1946. However, for years the ranch income and the telephone income remained too low to support the family. By necessity Howard worked a day job at Bridger Valley Electric for years while the children were young, and while he was employed there he also took care of the family business as a form of moonlighting.

When they were old enough, generation-three members cut their teeth in the field, sometimes taking on risky field projects. John G. described one of his first experiences with Union as follows:

> Our linemen placed two carrier huts, one at the Radosevitch ranch and the other at the location of the Lodore School. One of my first projects at Union was to help commission the carrier and connect the lines to provide service to the residents of the park. To do that, Raymond Tanner and myself would fly down to the dirt airstrip at the Radosevitch Ranch and use a truck that was stationed there to drive the line and connect the customers, place filters, and connect the carrier huts to the open wire. We were using a newly marketed open-wire carrier that was all transistorized. When we went to connect the Lodore School carrier, there was so much induced voltage on the line that you could draw an arc of twelve inches or more to ground. No one had thought to specify filters to divert the induced voltage to ground to keep it within limits. We were several days installing these filters using a short length of chain to short the conductors to ground as the wires had hazardous voltages on them. Once we had corrected this problem, it was possible to put the line into service.
>
> The airstrip at the Radosevitch Ranch was always a problem as the strip was west to east in orientation but sloped strongly to the west, hence you always had to land to the west and takeoff to the east. There was a power line at the east end to the strip; clearing that on takeoff could be problematical. I felt at times that we were so close that I could count the strands in the ACSR [aluminum conductor steel reinforced cable] as we flew over it. On hot days we often waited for the sun to go down before attempting the takeoff.

Generation three infused an abundance of talent and energy into the company at a time when it was getting its corporate house in order. Each of the generation-three members brought competencies and experiences to the company that helped position Union Telephone for a period of sustained growth, as well as technological modernization and innovation. The stage was set for the company to expand from the local level to the regional level.

TAKE ME TO YOUR LEADER

During its first century, Union Telephone has had to adapt its leadership structure to meet changing conditions. John D. had been content to stay above the corporate politics of the boardroom in order to stick to his main focus: to build the company. He later realized that this situation made it more difficult at times to do what was necessary to grow the business. Howard was the first Woody to serve on the board. It is no coincidence that it was only after he sat on the board that the company became profitable.

Union's leadership structure also changed throughout the years. The original bylaws of the company created the offices of president, vice president, secretary, and treasurer of the board, plus a manager position for the company. Howard first took over the presidency in the mid-1960s when Cliff Anderson retired, and he has served in that position ever since. John G. Woody describes the leadership-structure changes thusly: "Later, the bylaws were amended to replace the manager with the management team. Later still, [in 2009] the bylaws were amended to create the positions of chief executive officer, chief technical operations officer, chief customer-relations officer, chief administrative officer, and chief financial officer to replace the management team. Other titles were created and assigned by the board as seemed reasonable at the time."

A CRISIS

On June 1, 1979, one of Wyoming's all-too-frequent single-car accidents occurred on a lonely stretch of two-lane blacktop. This automobile accident involved John G. Woody and his soon-to-be second wife, Linda Kay Cox. John G., then serving the company as chief engineer, was seriously injured and came close to being a fatality, with grievous spinal injuries amongst his wounds. If he survived, there was a strong possibility that John G. would be a quadriplegic for the remainder of his days. Linda had an injured hand, and she lost part of a finger. John G. underwent long months of hospitalization and recovery. It was a tough time for such a close-knit family.

John G. Woody at the ranch.

At the time of the accident, Howard was still at the head of the company. It was difficult to lose John G. Woody's oversight of technical and operational matters, as well as his expertise in network development. The company continued moving forward, employing a management team to make decisions. While John G. was recovering, the management team consisted of Howard, Esther, Dee, Lee, and Jim. Esther and Dee exchanged shifts at work in order to maintain a presence by John G. Woody's side.

Generation-three member Jim naturally began to take on a portion of the operational decisions that usually fell to his brother John. In doing so, he expanded his job portfolio beyond his usual accounting and general management functions. During the next decade, Howard sought to reduce his personal participation in the telecommunications-trade organizations for which he had become a valued spokesperson. In 1989, Howard asked Jim to take over his board seat at the prestigious United States Telecommunications Association (USTA), the nation's premier trade association for providers of telecommunications services.

In order for Jim to assume this role and free Howard of the responsibility, it was necessary for Jim to have a leadership position in a telecommunications company. At that moment, Jim did not hold a title that would reflect the required leadership role. The family discussed the matter of a title for Jim, and, after careful consideration, they came together and agreed on the title of executive vice president. The title allowed Jim to assume Howard's position on the USTA board.[1]

The expediency of this appointment was regretted at a later date, but not due to any lack of ability on Jim's part. Howard was not aware that there were any feelings of competition between his children in terms of who would eventually rise to run Union Telephone once he retired. Union inadvertently had bumped into an issue that affects many family businesses: the succession of leadership.

Well over three-quarters of the businesses in America are family owned. These businesses eventually have to navigate the troubled waters of working out the order of succession for when family members leave the company, retire, or are incapacitated. What appears to be a simple business decision is not. Business decisions and family decisions are inexorably intertwined. To a certain degree, all of the politics inherent in rivalries and alliances between family members get played out on the job. The stakes tend to be higher than who mom or dad might favor the most. The manners in which conflicts are resolved affect the employees, as well as the customers.

When John G. recovered both his health and his mobility, he returned to work with the expectation that he would come back to the same situation he had left. Conversely, Jim had spent much time and effort towards guiding the company's operations, and he was hesitant to give up the authority that he had gained. John G. and Jim had opinions that were often at odds regarding the best operational direction for Union. The result was a company that had changed from a one-head/one-heart philosophy to a many-headed organization that was sometimes in conflict with itself. This was not an unusual situation for a family-owned business.

[1] *As it often happens when events happen rapidly and there is no record of the details, there are differing recollections of the principles regarding this event. Howard recalled the position as "general manager," and Jim as "executive vice president." John G. commented: "Jim made general manager well before he was on the USTA board. He was made executive vice president later, after he was on the USTA board and after the management team had come into being. This title had no express duties other than to represent UTC at the various industry-association meetings he attended."*

There were years with stormy seas ahead, but this same period corresponded with some of the company's greatest accomplishments. Like Homer's character Odysseus, the storms sent by an angry Poseidon did not end their journey, but instead served to make those who endured stronger and better able to take on new challenges. Some of Union's boldest advances emerged from the shadows of the chaos created by the succession issue. In true family fashion, the Woodys "cowboyed up" and focused on completing the task of building the corporation and sticking to the precepts of the founding dream. The Woody family was populated with hard-working, dedicated members who kept their noses to the grindstone no matter what. They could do nothing else.

Resolution of the succession issue was years off in the future. The process of determining a positive solution started in the 1990s when Howard read about some consultants who helped family-owned businesses with such issues. He requested that John G. contact Tom Davidow and Associates, who subsequently came to Mountain View and helped the family to work through the issues. Tom Davidow and Cindy Adams Harrison quickly learned about the family and the company. In order to gain as complete an understanding of the facts as possible, the consultants talked with each family member and company employee. Of great importance to a resolution was the consultants' ability to acquire commitments from the involved family members to be forthcoming in their views, and to also abide by the solutions that were generated through this process.

It took fourteen years for the balance in Union Telephone's management to be restored, but restored it was. During these years, John G. became the vice president of the company, and Jim regained his role as Union Telephone's spokesperson for matters concerning trade organizations and political committees. Today, their steadfastness, honor, and dedication to the company are never in question and are beyond reproach.

Howard always told his children, "The company comes first." It is another one of those phrases that could compete for being the Woody motto. This philosophy is a prime principle that drives each new stage of development for Union Telephone, and it has been ingrained in generations four and five. Out of the prolonged process of restoring balance in Union's management came a succession plan that will guide the company through the often difficult process of choosing between leadership options. The succession plan will help Union Telephone continue as a family-owned, family-operated business as long as there are family members who meet the standards.

It would be misleading to not acknowledge the significant role that was played by the generation-four members in redressing the structural difficulties in Union Telephone's management. Generation-four members had proven the utility of the new standards before these standards were codified. Generation-four member Eric Woody detailed the issues: "From the position of generation four, we played a very important role with taking the corporate structure from the management team to a more standard executive-officer structure that we have today. We as a generation became very concerned with not having the flexibility and strengths that we felt were necessary to compete in this industry. We felt that the management team had served its usefulness and was actually hurting the company as a whole." The fourth generation clearly demonstrated it was ready to lead.

The agreed-upon succession plan included the following prescriptions. In order to serve as an officer of the company, the coming generations are expected to meet the high standards that were set for generation four. All generation four and future board members are required to work in each of the company's departments, and they are expected to go to college and get a degree in a subject matter that is appropriate to the company's needs. Prospective board members are encouraged to hold a job outside of the company for a period of time in order to bring in experiences that will enrich the company. They then undertake the equivalent of an apprenticeship program where they work in each of the divisions of the company and learn the workings of the business. In addition to their predecessors mentoring them throughout the process, soon-to-be board members participate in (and complete) a formal training program. Even with all of these steps, there is an application process. The size of the board was limited to nine members, suggesting that board members must compete for the positions once there are more applicants than positions.

As Howard summed it up: "The Woodys are pretty protective about what they sign and agree to. I feel sure that these written declarations will surface again with generation five as they become available for future leadership decisions . . . once again by their choosing, not family pressure."

BREAKING UP IS HARD TO DO

Prior to 1982, AT&T provided almost all of the nation's long-distance telephone service, and it also controlled the Bell System, which provided most local service. AT&T's only serious competitor at the time was General Telephone and Electric (GTE). Additionally, AT&T controlled most of the manufacturers of telephonic equipment, essentially giving them almost complete control of telephonic communication in the United States. Some called this huge amalgamation of companies and subsidiaries the AT&T octopus, and others just called it Ma Bell. Whatever its nickname, the mega-company effectively shut out any and all competition. Even in the American Outback, AT&T controlled long-distance service. Ma Bell had not been interested in acquiring a remote telephone company like Union Telephone, but they still made money selling the company equipment and access to long-distance services.

In 1974, the US Department of Justice filed an antitrust suit against AT&T. The Kingsbury Commitment of 1913 (which was an out-of-court settlement of the government's antitrust challenge) had been a voluntary agreement and was unenforceable in court. The commitment did not keep AT&T clear of entanglements that led to a corporate monopoly of the telecommunications industry. Before the courts resolved the suit, AT&T agreed to the largest corporate breakup in American history, divesting itself of its many local corporations, but hanging on to much of its control of the American telephone system. AT&T proposed to break up the Bell system into nine regional companies, which became known as Baby Bells. A settlement was finalized in 1982 under the Reagan administration, and the breakup went into effect in 1984.

Divestment was big news, especially in the breaking up of the monopoly of telephonic-equipment manufacturing. Even so, Baby Bells (such as Southern Bell) continued to be major suppliers of telephone equipment. The FCC implemented a schedule of access charges for local companies to access and terminate a long-distance call. Because of this, local telephone rates tended to go up. Conversely, long-distance rates went down when companies such as MCI and Sprint entered the fray and gave AT&T stiff competition.

There still were great inequities in the system, especially for rural states like Wyoming. For example, the communities of Rock Springs and Green River are only twelve miles apart, but as late as the 1990s they were only connectable by a long-distance call and the toll charge that accompanied it. There is nothing but open range, a few manufactured homes, and a truck stop between the two communities and their outlying trailer parks. A local service area in a populous state like New York might encompass calls over twenty miles away and involve communities that collectively totaled hundreds of thousands of people. The long-distance providers (such as Mountain Bell, a Baby Bell) had little incentive to change such lucrative situations. The Kingsbury Commitment of 1913 had allowed AT&T to avoid antitrust action in the pre-World War I era by appearing to order the company to discard any further ambitions for a monopolized telephone system. However, appearances can be deceiving. The 1980s divestment settlement seemed to be a similar smoke screen, as AT&T remained in control of most of the national telecommunications system.

A DECADE OF CHANGE: EVERYTHING OLD IS NEW AGAIN

The 1980s was a decade when the direction of America was changing. Reaganomics, hostile takeovers, Medicare, mega-mergers, AIDs, Pac-Man, an extra-terrestrial named E. T., rap music, the Space Shuttle *Challenger*, the Human Genome Project, and the end of the Cold War were all hallmarks of the decade. Society called the growing number of status seekers who were coming of age the "Me" generation.

The Union Telephone central office with the familiar telephone booth.

For Union Telephone, the 1980s were a time of continued growth and modernization. Union Telephone installed its first digital switches in 1980. This was only the second such switch installed in the entire state of Wyoming. Roger Lind was promoted to outside plant superintendent in 1980, and John G. was appointed assistant manager in 1982. The company settled into its new headquarters location in Mountain View in 1983. Throughout 1985 and 1986, digital micro-wave systems were installed to replace the analog systems. Also, Union continued its decades-long process of replacing open-wire lines with the plowed-in lines, allowing them to retire the Lonetree–Mountain View open-wire line.

Roger Lind plowing in telephone lines, replacing open-wire lines with more efficient coax and fiber lines.

OPERATOR, CAN YOU HELP ME MAKE A CALL?

In 1985, at a time when many phone systems were automating customer services such as pay phones, directory assistance, and operator-assisted calling, Union Telephone returned to the days of live operators. The return of operator service required some of Union's classic innovations. The manufacturer of the Union operators' board had failed to provide software for rating and billing, so Jim Woody devised software for these purposes. After successful tests of the system, Union was ready to provide these services.

Having real people answer the phones to assist customers with their service needs was another example of attempting to maintain a personal touch in an industry that was rapidly becoming impersonal. It was a risky return to something from "the old days," especially in an industry where features common just a few years prior had already passed into obsolescence. It was a bold move to put computers out of work by hiring people to do the work, although the operators soon began working with computers to provide customers with the best features of both worlds.

The operators had a unique public-service job. At one moment they were asked to help a traveler connect with a transportation provider; at another moment they received a distraught call from a young woman being held captive as she tried to call a relative for help. Supervisory Operator Joyce Garris summed up the advice she gave all new operators: "You never know what it's going to be when you pick up the phone and open that line. If it's an emergency and the panic is there on that end, don't go with them. Wait until you get things done, hang up, and *then* go fall apart."

The operators worked long hours. There were usually at least two operators available on a shift. The slow times were filled with conversation, Scrabble, card playing (two of the operators created their own personal version of the popular card game Rummy), and jigsaw puzzles. All of the Union operators were part-time workers, but they formed an effective team, covering for each other when personal time and vacations came up.

Many of the Union Telephone operators had experience in the profession before joining the company's family. Union Telephone's pioneering move to computerize the operators' consoles required a bit of adjustment, even for experienced Hello Girls. Some of the workers had no experience with computers and had to supplement on-the-job training with courses at the local community college. Once they had learned how to operate the new consoles, the operators liked the convenience of the new technology, but for a while some of them had their doubts about their ability to keep up with the computer age.

In the days before credit-card calls, phone cards, etc., one dropped coins into fitted slots at the top of the pay phone. The chiming created by a nickel dropped in the coin slot was different than that of either a quarter or a dime. Prior to the days before automated counting, the operator had to count the varying chimes (and later on, tones), hopefully identifying all chimes correctly to ensure that the call was paid for before being put through.

Operators commented about hearing those coin tones so much that the sounds haunted their sleep. The operators quickly became proficient in recognizing the tones and tallying up the amounts deposited. The seasoned operators were difficult to fool. Throwing a penny down the nickel slot usually jammed up the phone. Some larcenous customers created fake coins—called "slugs"—to put in the slots to fool the machine. Others tried to defeat the system by playing back recordings of the noises made by coins as they were dropped into the slots. When all else failed, the customers sometimes just argued with the operator as to whether they actually put in sufficient funds. There's something to be said about automatized pay phones: it is difficult to argue with a machine.

UNION CABLE TELEVISION

In 1986, Union acquired the Mountain View Cable system from Angwin Cable Television, and for a few years they were the provider of both telephone and cable services for much of the Bridger Valley. This acquisition was prompted by the retirement of the cable system's owner, as well as a desire to maintain service for the local community. For Union Telephone, cable television turned out to be a lot more trouble than they had bargained for. As satellite firms began dominating the market, Union allowed the cable system to follow its former owner into retirement, and the service was discontinued in 2011.

RETURN TO BROWN'S PARK

In 1988, a Union Telephone crew reworked the Dutch John–Brown's Park telephone line. Union was responding to a request to provide service to a pumping station in the Clay Basin, a geologic feature that mainly serves as underground storage for extracted gas waiting to be shipped to market. The pump station in question served a phosphate-slurry line between the greater Vernal area and the J. R. Simplot fertilizer plant south of Rock Springs. The client required voice and data services for a pumping station. There were many difficulties regarding engineering and access to the project site, but the customer agreed to foot most of the bill for the creative solutions required.

A twelve-pair T-screen cable replaced the open-wire system. The United States Forest Service refused access through their lands to the project site, but Union Telephone was able to get permission for access along an abandoned pipeline for the segments between Dutch John and Little Hole, which required the innovative use of a self-propelled lawnmower to pull the cable through the pipe. Helicopters were used to deploy the telephone line in the areas around Brown's Park that lacked roads. It was by no means an easy path for admittance into one of the most isolated corners of America's Outback.

The Union crew included Howard, Roger Lind, Kim Halford, Ken King, and Chuck Fagnant. This small group had the adventure of a lifetime as they worked within a landscape that was best suited to mountain sheep. They traveled from their remote campsite to the work areas on horseback. The lines, being too heavy for horses to haul, were brought to the project area by helicopter. The horses were utilized to pull the heavy line into position. The project was physically

Roger Lind using a horse to pull a telephone line into position in order to provide service to the Brown's Park, Colorado, area. Roger eventually became the outside plant manager for the company until his retirement.

demanding for both the men and their horses, but Union Telephone employees returned with memories that money just could not buy.

Once the line to Brown's Park was completed, the remainder of the line was plowed into the ground in the conventional fashion. However, even this part of the project required some adjustments. The local Bureau of Land Management office would not approve the most direct route through Jesse Ewing Canyon, so the line was plowed in through Red Creek Canyon to the Clay Basin. Years later, this same line was exposed by a flash flood and required reburying.

The crew on this unusual project was required to "rough it" both during and after work. Construction crews of the time usually expected hotel accommodations with restaurants providing the meals. The Brown's Park project did not allow for such niceties. A camp consisting of a few crude trailers was established near the work areas. The mobile homes were hauled over the twisting, bumpy trails that passed for roads in that isolated area. This was somewhat of an adventure in itself.

The participants took turns preparing food for each other, a situation that made for an occasional meal so poorly prepared that participants still talk about it decades later. Camping together and working together builds a sense of camaraderie. The crew finished their work feeling both pride in the accomplishment and a twinge of regret that the adventure was over.

YOU MUST BE NEW IN TOWN

As the industry continued to increase in technological complexity, it became difficult for Union Telephone to add staff members who possessed the proper skill set without having to seek employees from outside the Bridger Valley. Even by the standards of America's Outback, Mountain View is a small town, and it had just over one thousand inhabitants in the early years of the twenty-first century. Although there were another two thousand souls in the nearby town of Lyman, there were just not a lot of people of employable age in the Bridger Valley job market. Competing for qualified labor spurred Union Telephone to improve its employee benefits, including retirement packages and the usual amenities.

Even so, new employees sometimes were not a good fit with the local community. Employees recounted one tale of a brash New Yorker who came to work in Mountain View. The man was technologically astute and could do the job well, but what might pass for being a regular guy in New York City comes off as a bit abrasive in rural Wyoming. One day, a couple of his coworkers decided to play a trick on the man. When the New Yorker was high up a pole, they piled some tumbleweeds around the base of the pole and set fire to the tangle of brush. Tumbleweed does not burn very hot, and the danger to the pole climber was more perceived than real. The startled New Yorker did not approve of being what the locals sometimes call "hoorahed," and he left the company shortly thereafter. This was not typical treatment for an outsider, but it exemplifies one of the problems in bringing in and retaining talent from outside the region.

Many talented individuals have difficulty adapting to the rural nature of the Union Telephone service area. In small communities like Mountain View, newcomers sometimes find the isolation, harsh climate, and provincial cultural climate a struggle to overcome. Even so, the advantages of clean air, clean water, stunning vistas, availability of land, low crime, and a hard-working and tight-knit community are powerful attractants for the technical talent necessary for a communications company to compete. Throughout the years, Union Telephone has done well in building capabilities from within and—when necessary—recruiting talent from outside.

YOU'RE NOBODY 'TIL SOMEBODY SUES YOU

Throughout the years, Union Telephone has had a love-hate relationship with AT&T. At times, the two companies competed against each other, but in different circumstances they cooperated with one another. In 1988, AT&T filed a complaint with the Public Service Commission that disputed Union Telephone's entry into the toll-service area. AT&T followed this up by filing a suit against Union Telephone for an antitrust issue. The matter between the two litigants was settled out of court. It must have seemed ironic to be sued by the company that only six years earlier managed to avoid the federal government's antitrust litigation by self-imposing the largest corporate breakup in history.

By the 1990s, relations between the two companies had warmed sufficiently such that AT&T sold Union Telephone the cell licenses that the FCC was about to reclaim.[2] Union Telephone made arrangements with its competitor for roaming agreements, whereby customers from a competitor could use Union Telephone facilities. Union Telephone's system was depicted on the AT&T coverage map that was used in their advertising. However, the negotiations apparently were not completely smooth, as AT&T Wireless wanted a lower per-minute rate for roaming charges. John G. knew he had to convince the Union leadership to take the deal because the AT&T Wireless roaming agreement promised to result in a significant increase in volume. This matter is discussed in more detail below.

UNION TELEPHONE BECOMES UNION CELLULAR AND THEN UNION WIRELESS

Generation three took Union's helm during one of the corporation's greatest gambles, which subsequently became one of its greatest triumphs: the jump into cellular telecommunications. Howard had laid the groundwork for Union Telephone's entry into this new technology with his previous installation of microwave technology and radiotelephones. These gave Union Telephone a leg up when testing the waters of the cellular-communications industry. Even with these advantages, the decision to jump into the cellular market was not an easy one. The board members were divided into two camps: those who felt the new industry was too risky, and those who saw a golden opportunity. The board wrestled with this decision for some time, but in April of 1989, the decision was made to roll the dice. The cellular system was a go.

Union's old partner, the REA, funded a five-million-dollar loan, helping to make the company's jump to cellular possible. These funds, in addition to

[2] *John G. indicates that the frequencies were purchased from AT&T Wireless of Bellevue, Washington, which had been Macaw Cellular and was later, in turn, purchased by Cingular and renamed AT&T Wireless.*

Union's profits and a loan from CoBank, were used to build the cellular service. The future of the company rested upon the outcome of this leap of faith.

Kim Halford working on one of several towers placed within the greater Jackson Hole area. Kim eventually became the microwave cellular maintenance director until his retirement in 2014.

Cellular technology is a form of radio communication in which stationary base stations or towers are spread across the landscape at intervals. Because of the ability of the stations and towers to re-use frequencies (an attribute that allows multiple linkages to use the same tower), cellular service offered unprecedented mobility. However, in the early days, it was not at all clear that the new technology was going to be profitable. The great expense of creating a cellular network might not have been recouped if the technology turned out to be a fad. Unlike landlines, there was no cost-plus guarantee regarding the expenses. If this gamble was to pay off, Union had to build their system and contend within a highly competitive market against the largest telecommunications companies. There was a real chance that Union Cellular might build a few towers that handled so little traffic that selling them to a competitor or abandoning them became necessary.

Cellular networks normally have towers arranged in a pattern that is designed for efficient coverage. The ideal situation was to configure the tower pattern so that mobile devices maintained contact in a fashion that consumed less power. Adding stations and towers to the system effectively expanded the horizon and extended the networks. In theory, a patterned grid of cell towers in a relatively flat landscape provided reasonably seamless service. The flattest places in Wyoming tended to be atop table mesas. In order to provide coverage with maximum efficiency, the engineers had to become adept at working within the constraints imposed upon them by the terrain. Even with the best planning, Wyoming cellular service was fraught with "bad air," a colloquial name for poor-reception areas where the terrain conspired to create zones in which signals were blocked. It is a wonder of engineering how the stations and towers balance the use of mobile devices and are able to pass a transmission from one station to another as the customer moves through the complex landscape of America's Outback.

There were numerous systems available for handling the connections to multiple moving devices while providing clear and uninterrupted service, such as FDMA (Frequency Division Multiple Access), TDMA (Time Division Multiple Access), PDMA (Polarization Division Multiple Access), and CDMA (Code Division Multiple Access). In the 1990s, the practical application of cellular technology was new. Within twenty short years of their introduction, cellular systems encompassed much of the globe, providing access to telephone, messaging, and Internet services.

If one asks an engineer about the technologies used by Union in developing its cellular system, the explanation quickly becomes a spiderweb of complex telecommunications acronyms. According to John G., the original Union cellular system was analog, in which voice signals were transmitted as FM (frequency modulated) signals. The only data stream on this system was the control between channels, which set up and controlled the call and allowed hand-offs between towers. The digital systems CDMA and TDMA were used by Union in the mid-1990s. In the late 1990s, Union transitioned to GSM (Global Standard for Mobile), which is a variation of TDMA that offers increased spectral efficiency. In its centennial year, Union was using UMTS (Universal Mobile Telephone Service), a CDMA technology with improved efficiencies. Union is in the beginning stages of deploying LTE (Long-Term Evolution) through OFDMA (Orthogonal Frequency Division Multiple Access), which further enhances spectral efficiency and speed.

Most of the major companies in cellular communications initially built their systems where there was already substantial conventional phone service. The majors focused on populated areas that contained higher densities of landlines. At the time, travelers had access to a vibrant network of pay telephones. Accordingly, in the early days of cellular, mobile service was not necessarily a hot item that was guaranteed to turn a profit. Mobile service at that time was a utility choice of corporate customers and the well-to-do. By embracing rural cellular service, Union broke the mold created by its competitors. It was a gamble that paid off.

Since cellular systems are, in essence, a complex system of radio transmitters and receivers, the frequencies available are very important. To be a player in the coming cellular world, Union Telephone had to establish a territory for their system. The invisible and intangible world of radio frequencies is a magical kingdom where bandwidth is won in lotteries, traded, bought, and sold much like land. Union Telephone used the 1988 and 1989 lotteries, some plain old horse-trading, and partnerships with both US West and Kamas Telephone to acquire bandwidth in territories in and around their landline services in Utah, Colorado, and Wyoming. Bandwidth is a system of intangible telecommunications lines waiting for appropriate transmitters and receivers to use it. Once a company received bandwidth in a geographic area, the game became very much like homesteading the airwaves, where territory was allotted based upon a company's performance at "proving up" in a specified amount of time. The race was on.

As John G. explained: "[The acquisition of these rights] entitled you to get your construction permits to build out a system. You had five years to cover as much as you could. Because that is what you get to keep." Companies that received bandwidth from the Federal Communications Commission had eighteen months to get their first tower up. Given the unpredictable weather and complex land-ownership patterns of Wyoming, the completion of the first tower was not a sure thing. Most of the towers erected by Union were thirty-foot-tall lattice towers, though circumstances sometimes called for fifty or eighty footers.

Many of the sites used by Union required extensive environmental analysis and consideration for the fragile landscapes in which the prospective tower was built. For the first time, Union experienced a slowdown in prospective construction due to the complex ownership of land and the various hurdles created by the landowners and managing agencies. John G. was most proud of the tower the company built on National Park Service land on Signal Mountain, near Jackson Lake, Wyoming. "It took us six years," he recalled. The selling point to the National Park Service revolved around an aboveground power line that created a visual intrusion in the landscape. Union offered to plow the above-ground line into the ground, thus greatly reducing its visual impact. The park approved of this trade-off and permitted Union to build the tower.

This was not the only time that Union dealt with a landowner who had concerns about how their site would look. On Bakers Peak (located in northern Colorado), the company wanted to put a tower near a landowner's cabin. They had to build a site that looked much like a cabin, with the tower rising from within the structure. The unusual base for the tower appears to have pleased the landowner, but it accomplished little to mask the presence of the tower.

The Federal Communications Commission used the locations built during the "homesteading" of the airwaves to draw hypothetical coverage contours around each of the towers built, effectively creating a territory. When the Federal Communications Commission granted the use of radio frequencies in December of 1988, it allowed the company to become a player in the expanding world of cellular telephone service without having to lease the frequencies from competitors. Union followed its success in the lotteries in 1989 by acquiring additional bandwidth in Colorado in the vicinity of the Interstate 40 corridor. In July of 1990, Union Telephone officially opened Union Cellular service with a gala event in Mountain View. The company had only built a modest eight towers with which to begin providing cellular service, but these rapidly multiplied. In its centennial year, Union boasted over three hundred towers in service.

Union Telephone celebrated the kickoff for its new cellular service in 1998 with employees and the community alike. The telephones did get smaller.

When building its cellular network, Union Cellular chose a model different from the larger companies. They started building cellular stations in unpopulated areas, sometimes in the middle of nowhere. Mountaintops, such as Elk Mountain, Cedar Mountain, Juniper Peak, Thunderhead, Aspen Mountain, the Winter Park Ski area, and other prominences allowed Union to maximize the size of their fledgling system's coverage area. The company had significant experience in working in these remote areas, but there was still a learning curve to building the cellular towers. There was no carefully considered master plan, but the company remained flexible in order to rapidly seize opportunities as they arose. John G. described Union's process during the initial stages of building the cellular network as "mountain hopping."

Union Crews have become so adept at stacking iron that cell towers can be built in a few days in remote locations.

Experience had shown Union that even an isolated patch of America's Outback might suddenly spring to life with the discovery of gas, oil, uranium, coal, lithium, or even the building of a dam. Union was jumping into the cellular poker game holding decent cards. The gas fields around La Barge, Pinedale, and Kemmerer were in the early stages of becoming the busiest gas prospects in North America. Union Telephone already was building radio towers in the "gas patch" during the time when places like the Jonah Field, the Anticline, the Mesa, Opal, and La Barge were about to become some of the most sought-after drilling sites in the United States.

Drilling rigs, compressor stations, pipeline facilities, worker camps, offices, sour-gas treatment plants, and equipment yards all had communications requirements, as did the tens of thousands of workers in construction, maintenance, surveying, and seismic exploration who came from all over the world to get in on the boom. Along with them came a swarm of what the military used to call "camp followers," the people who make their living prospecting the prospectors. Keeping the well-paid work crews entertained and fed required thousands of additional workers. Wyoming's infrastructure is normally stretched beyond its limits during the early stages of a boom. Local communities were challenged to meet the expanding needs for social, health, and emergency services.

A man camp might spring up almost overnight in a deserted valley, suddenly creating as much demand for telephone service as a small Wyoming town. A large pipeline-compressor complex might have all of the communications needs of a small company. The numerous construction crews, maintenance people, regulators, surveyors, and others were likely to need the service—miles from the nearest town—for personal use, up on remote mesa tops, or down in deep canyons.

IF YOU BUILD IT, THEY WILL COME

One of the distinctions that sets Union apart from its competitors relates to the construction of its cellular towers. Most people have noticed cellular towers in skylines. One is tempted to view these structures as antennae but not much else. Each cellular tower is a technological complex unto itself. These sites are not exact copies of one another, and they have variations in their configurations depending on the particular needs at each location. All of the construction work and installation of the Union tower sites was performed by Union Telephone employees. This gave Union a competitive advantage in constructing their network. Had Union Telephone chosen to contract the work to build the towers, it would have cost the company roughly three or four million dollars for each one of the larger towers. Using tall topographic features such as platforms allowed Union to build shorter towers. As their internal expertise in designing and building tower complexes grew, Union cut the cost down to thousands of dollars per tower.

While the company was initially learning the ropes, erecting a tower was a slow process that took weeks. However, as the crews' expertise peaked (coupled with a systematic approach designed by Eric Woody), it became possible to erect a facility in not much over a week. Eric emphasizes that the redesign of their procedures was driven by safety first and efficiency second. The construction process has become so refined that one crew was able to stack an eighty-foot tower on its cement pad in a single day.

The learning curve experienced by Union personnel served the company well. Generation-four member Eric Woody designed a vehicle that was instrumental in raising the shorter towers. Once again, Union supported innovations that were created within the company. With the low building cost of the towers and the speed of construction, Union was able to put seventy-four new towers into service in the span of a single year.

As Union Cellular started installing its new cell towers, something unexpected happened. When a tower came online, it almost immediately began to handle calls from individuals who were subscribers with other companies. This proved to Union management that there was an opportunity to establish roaming deals with these other providers. Roaming charges were easy money for cellular companies. They did not have to service the customers but were able to receive time charges from the other providers. Soon, competitors in Rock Springs were offering "A coverage" and "B coverage," the latter being a Union Cellular roaming charge of seventy cents per minute. US West set up a roaming agreement as well. As the gas boom continued to flourish throughout the 1990s and into the new millennium, these agreements established a substantial revenue stream for Union. In the early years of Union's entry into cellular service, roughly half of Union's revenue came from roaming charges. Jim Woody remembered: "We went from zero to a million dollars a month in a period of about four months. For us it was like a Godsend, we had dollars we had never seen before."

LITTLE HOUSE ON THE PRAIRIE

At each cell-tower site there is a small building to house the complex electronics that are the nerve center of the site. The electrical devices inside require climate control, such as air conditioning in the summer. Union uses concrete pads as footings for its towers. The towers are fully grounded to protect them from lightning strikes. Usually there are one or two microwave dishes attached to the latticework tower of each unit. Once a site has been selected and approved, it takes anywhere from a few days to a month to build the complex and bring it into operation.

In Wyoming, buildings that lack sturdy fencing around them quickly become favorite scratching posts for cattle. Fences have to be high enough to keep deer from jumping over them, and sturdy enough to withstand challenges by wild horses, cattle, sheep, deer, elk, moose, buffalo, and even bears. All of the equipment at each cellular site is linked to the power grid by an underground power line. In many instances, this requires laying miles of power cable to the site. There is also an access road, which for many towers is a two-track path worn into the soil, but for most it is a graveled road.

TOP O' THE WORLD

Cell towers are often erected upon some of the most exposed terrain on the planet. The Union cell tower on Elk Mountain in Wyoming (estimated at roughly 11,110 feet above mean sea level) is regularly pummeled by winds exceeding 120 miles per hour with temperatures that can dip to below minus forty degrees.

It was at places such as Elk Mountain that we learn just how brittle our technology has become. Long-term Union employee Kim Halford described the difficulties with maintaining the Elk Mountain tower. "The environmental conditions are kind of unique. It may take from four hours to two days to get in through the snows. We had a hard time keeping an antenna. We had to beef up one part after another. When a fix is put in you find out where the next weakest part is."

Technician Vance Walker finishing the installation of a facility atop a mountain.

Mother Nature had more tricks up her sleeve than just high winds, sleet, and snow to keep the technicians busy. Generation-four member Eric Woody remembered going to a site and working in the equipment structure. When he entered the structure, the temperature was between fifty and sixty degrees, but as he recollected, "When I came out of the shelter it was below freezing with whiteout conditions."

Animals keep things lively for the tower crews and maintenance technicians. Ravens, a species protected under an international treaty, like the dishes on the towers and often poke their talons through the protective covering. They also eat the insulation on the wiring. This forces the crews to install bird spikes on both the wiring and the tower to discourage the birds. Bird spikes are stainless steel or plastic rods that vary from two to four feet in length and are arranged in a spread array, much like a hedgehog's back. The bird spikes are attached to surfaces where a company wants to prevent birds from landing, roosting, or nesting. The down side of spiking a structure is that installing a proper array of bird spikes makes servicing and maintaining the structures all the more difficult.

Snowcats are often necessary equipment for tower work in Wyoming. This snowcat in the Jackson, Wyoming, area is beneath a precarious shelf of snow that is typical of the mountains. When Snowcats are not available, the work site has to be entered by snow shoe or other snow machines.

Cellular towers have had infestations of wasps, marmots, mice, and rats. One time, a technician working in a complex's building left the door open. Turning around, he saw he was not alone. A young black bear was standing in the doorway, curiously sniffing the air and peering inside while trying to assess what was occurring within the interior of the structure. Startled by the sudden appearance of the bear, the technician panicked and stampeded out the door, running over the inquisitive bear in the process. The technician is reputed to have said, "I'm not sure who was more scared, me or the bear." We may never know, but the bear never returned to the site.

Driving to and from the cell towers is often an adventure in itself. Mud, snow, whiteouts, snowdrifts, flat tires, dust storms, fallen trees, mechanical problems, livestock drives, livestock, wildlife, electrical storms, tornadoes, stuck vehicles, locked gates, and floods can all conspire to add complexity and challenges to any trip. Simple creatures like antelope often make driving difficult. Antelope have a habit of running in front of whatever they're fleeing from, making sharp turns that are intended to break a natural predator's stride. This defensive instinct only adds to the danger of being hit by a vehicle. Wild horses might

Cell towers endure extreme weather conditions.

take offense to the presence of unwanted guests in their territory and aggressively charge vehicles or survey equipment. Moose, bears, and elk all have their own rules of engagement. Bounding deer can appear in the headlights as if by magic from out of the gray of the twilight almost anywhere in the backcountry.

Generation-four member Brian Woody recalled having to rescue a crew that had become snow-bound while trying to work on Elk Mountain. A track had slipped off of their snowcat, and the rookie crew could not repair the machine. During the early afternoon of a snowy day, Brian received the call for assistance while working in the Lander area. Brian "drove by braille" for miles, passing through numerous ground blizzards. Leaving the highway several hours later, the rescuers negotiated an unplowed road with frequent drifts. They managed to push through most of the drifts, but when the giant accumulations of snow held their truck fast, they unloaded their snowcat and pulled the truck through, always moving towards the crew on Elk Mountain. This happened several times before they reached the base of Elk Mountain, making the journey seven and one-half hours at that point. The snowcat was unloaded in the dark, and the cross-country part of the story began.

A second rescue crew (this one from Rawlins) had reached the mountain about a half hour before Brian. That crew had inexplicably slipped deep into the caldera and was likely to remain there until spring if they continued down much further. Brian picked up five people stranded in the broken snowcat and then went back to the other rescuers. The rescue crew had buried their snowcat up to the flashing light atop the cab. It took a great deal of time and effort, but after extracting the stuck snowcat, all personnel were evacuated to Rawlins. It had taken twelve hours to get this much done. Brian and the crews returned to Elk Mountain the next day and repaired the disabled snowcat. After four hours of sleep, the workers were assembled for their return to Elk Mountain. The adventurous part of the incident was over, but there was plenty of work yet to be accomplished.

In order to be effective in their work, the training for tower-maintenance crews is intensive by necessity. The crew members have to know arctic survival techniques, first aid, automobile repair, map reading, navigation, lineman skills, wildlife behavior, air-conditioning maintenance, house and roofing repair, and off-road driving skills. A sense of humor and an abundance of common sense are among the most useful tools. In addition to possessing all of these skills and qualities, the crew members are expected to know the electrical-technician skills necessary to maintain the complex electronics at the tower, and be able to climb the towers in high winds and other dangerous conditions. Kim Halford described these special technicians as "Union's SEAL Team." Supporting Mr. Halford's assertion, Matthew Myers of Union was the winner of the 2011 "Toughest Site Competition" held by global communications provider Anritsu. The award recognized the exceptionally difficult conditions Myers and his team routinely encounter while maintaining roughly forty cell towers, one as high as 13,000 feet above mean sea level.

IT PAYS TO ADVERTISE

When cellular service was new to consumers, it took more than a little salesmanship to convince the public to adopt the new technology. Originally, the mobile units were roughly two thousand dollar in price. As such, they initially remained a service for the well-to-do. As the service costs dropped to an average of roughly two hundred dollars, consumers adopted it more readily. Union held a series of town meetings and demonstrations to show their clientele what cellular service could do for them. Family members like John G., Pamela, and Jim (as well as trusted employees such as Roger Lind) traveled to many of Union's service communities hoping to drum up business for the new system. It worked.

Along with the cellular telephone network came a new approach to advertising and customer outreach. Telephone companies like Union were used to being somewhat of a local monopoly. When someone moved into a community, there were usually only one or two available carriers for landline telephone service. Cellular service changed the whole game, and it became more critical to offer competitive services.

Before their entry into the world of cellular communications, Union Telephone did little mass-media advertising. However, Union Cellular jumped into the world of advertising with both feet. At first, Union focused on radio and a little newsprint advertising. As Jim Woody recalled: "People didn't think we were a real company like AT&T. You just can't spend money like they spent." Even so, the word about Union got around, and the demand for their cellular service grew rapidly.

In the cellular era, a typical duty for a corporate CEO is to be the face of the company. Howard has starred in several commercials, and his face has graced billboards and the Union website.

The company's advertising has featured Howard in television spots and billboards, which has given Union a "CEO personality face," a concept that is popular in Madison Avenue thinking. Howard's winning smile and folksy way of talking have lent themselves perfectly for an advertising campaign in America's Outback. Here was a face and a voice for the company that was familiar, neighborly, and trustworthy. After fifteen years as Union Cellular, the name was changed to Union Wireless in 2005. The emphasis on mass-media advertising that was begun under Union Cellular has continued unabated.

The company opened storefronts in its service-area communities where future customers could shop for their phones and service plans. It was a sound marketing strategy to get out into the communities, especially those where the extended energy boom brought in a steady stream of new users with high-paying jobs.

BEGINNINGS AND ENDINGS

The advent of cellular service marked the 1990s as a major milestone for the company. However, cellular service was not the only noteworthy event for the company. In 1991, John G. became the vice president of the company and was elected to the board of directors. In 1992, Howard decided it was time to end his long run as general manager of the company. He officially became "semi-retired" at that time. A year later, Jim was appointed general manager in his place. In

1993, the company sought exchanges in Elk Mountain, LaBarge, Hanna, Rock River, Encampment, Shirley Basin, and Afton. In 1994, Union borrowed $30.5 million to purchase seven exchanges from US West, one of Union's competitors.

In1995, Union amended its corporate bylaws and established a management team to lead the company, which eliminated the positions of manager and vice president. The initial management team consisted of Howard (president), Esther (treasurer), John G. (chief engineer), Jim (research and corporate development), Dee (operations), and Lee (central office).

As the world entered a new millennium, there were still four members of generation three who were not on the board. This state of affairs did not long endure. Jim and Dee were added to the board of directors in 2001, and Lee joined the board in 2003.

HUMAN RESOURCES

Growth imparts changes other than size. In the early days, there were few employees. Union did not reach twenty-five employees until 1990, seventy-six years after the company's founding. With the onset of the cellular boom, the days of slow growth were over. Union's staff doubled by 1995, and it doubled again by 1998. Union Telephone had two hundred employees in 2004, and at the time of its centennial celebration that number had grown to 271. In its first one hundred years, it is estimated that Union Telephone employed a total of 954 individuals.

As the company added more employees (such as retailers, operators, public-relations specialists, and positions necessary for the construction and maintenance of the burgeoning cellular system), employee relations came to the forefront of Union's practices. The company has a long track record of trying to treat its employees well. Employee benefits had been in existence as far back as 1969, when the first pension plan was established, and 1972, when the initial attempts were made to codify employee relations and work schedules. Wages were among the best in the area, and even part-time employees had

Esther P. Woody

a benefits package. Recognizing the dangers inherent in the region's winter-driving conditions, the company once provided bag phones for employees to keep in their automobiles for use in an emergency.

The company also established a Human Resources department. Originally, Esther headed this effort, but she eventually brought on a full-time person to help her. The Human Resources department established rules and training for personnel issues and safety that were comparable to those in other Wyoming industries. The department created an employee handbook in 1996, and in late 1997 a 401K retirement plan for Union employees was established. The board approved an employee compensation policy in 1998.

GOD IS MY COPILOT

In January of 1997, Jim boarded his plane, intending to fly himself home from a meeting in Price, Utah. As Jim completed the final instrumentation checks, the external temperature plunged to roughly minus fifteen degrees. Flying in such frigid weather conditions had become second nature to Jim, and they were more of a predictable inconvenience than an increase in danger. The controls were a little sluggish as the plane climbed, but this was nothing unusual.

The cloud cover that day was a broken bank of clouds at roughly 12,000 feet, so Jim piloted the plane to 12,500 feet to get above the cloud bank. The steadily droning engine grew louder as the plane strained to gain altitude. As the craft neared Heber, Utah, the plane shuddered as the engine began to shake. Jim checked the instrument panel, but nothing was registering as being amiss. As he reduced power to the engine, the shaking stopped. It was a puzzling situation.

At that time, the plane was roughly fourteen miles from Heber, a community that had a small airport. If Jim had proceeded on his original flight plan and been forced to land in the wide expanses of Utah, he would have been fortunate to land in one piece. Once on the ground, it would have been a miracle if assistance could have reached him quickly in the sub-zero temperatures.

Something told Jim it was time to head to Heber. He immediately changed the direction of the ailing plane and began his landing approach. During his final approach, the oil pressure gage dropped to a value of zero, and the plane began to "bleed out" air speed. He managed to get the plane safely on the ground with the last of the plane's kinetic energy.

Jim Woody at the ranch preparing inoculations for the family's cattle herd. *Photograph courtesy of Linda Montoya*

Safely on the ground, Jim pulled the cowling and discovered that one of the plane's engine cylinders had broken through the intake manifold. The shaking caused by this malfunction had loosened many of the bolts that held the engine in place. By following that inner voice and making that split-second decision, Jim had avoided an almost certain fatal crash.

Jim, who is a deacon in his church, explained: "God was looking after me that day. If I had gone to Price, I would have been above the overcast, and if I landed would have frozen to death as the route was unsettled." While the event clearly was unsettling to Jim, he remains an avid flyer, and he even built his own plane. He does not view flying with fear, as he knows he can count on help from his co-pilot when it is needed.

A REMARKABLE GENERATION

Generation three had bridged the gap from the twentieth century into the new millennium. They had earned their stripes during the boom period of the seventies, and at the time of Union's centennial, generation three still provided solid direction for the company. Like the generations before them, generation-three members well understood the mechanics of constructing and maintaining the technology in the field. It is amazing how much of the technological transformation was performed directly by members of generation three. Rather than rely upon a string of outside experts, generation three mastered the complexities of new each new technology as it was developed and made available. The generation's performance suggests that one of the Woody family mottos could be: "Never stop learning the ropes."

Under generation three, Union Telephone was transformed from a profitable system of exchanges servicing southwestern Wyoming and adjacent valleys in Utah and Colorado, into a regional competitor whose service area encompasses much of Wyoming, large portions of Colorado, and significant areas in Utah as well. The wireless extended network reaches into distant places throughout the nation, including Alaska. The leap into cellular was a great success, and by carefully acquiring bandwidth (including some not in use by the technology of the time), Union Telephone positioned itself for their next chapter: dancing technology. Our next connection takes us deeper into the American Outback and brings Union to other continents.

UNION TELEPHONE
Union
100 Year
Anniversary
1914 - 2014

CELEBRATION

You and your guest are cordially invited to a special celebration commemorating Union Telephone's 100th Year in business.

Please join us on:

DATE

Tuesday, January 28th, 2014

TIME

Social Hour 6:00 pm - Dinner 7:00 pm

LOCATION

Depot Square
920 Front Street - Evanston, Wyoming

Please RSVP by Tuesday, January 21st, 2014 to Ruth Barclay - (307) 782-4169, on-line at: www.greenvelope.com/event/evanston or email at: 100YearCelebration@UnionWireless.com

Chapter 5

The Fourth Generation: Technological Horizons Within a Changing Industrial Landscape

Nothing endures but change.

–Heraclitus

In its centennial year, it was generation four that dominated the management positions and the board of directors. As the fourth generation of family members came of age, the company was already well established and profitable. The members of previous generations wondered: With the struggles of the past already overcome, how would generation four move the company forward?

It was difficult for Union's elder generations to look into that hazy horizon of the future and foresee what the next generations might deal with. The company had already survived the Great Depression, two world wars, numerous "police actions," a cold war, internal strife, minerals booms, minerals busts, several recessions, and found itself in the middle of a global war on terrorists, where the government was paying close attention to internal communications like never before. Whereas in the early years the Union board of directors might think exclusively of local concerns, it eventually became necessary to focus upon regional issues when the company became a regional service provider. As Union's horizons and influences expanded, the challenge for the fourth generation is to envision the company in a global economy with international politics while still maintaining a watch on more traditional local and regional concerns.

CHANGING OF THE GUARD

Union Telephone Company exists both because of and for the family. The company has developed because of the decades-long efforts of Woody family generations and their loyal employees, all of whom work within the ethical framework and business model established by John D. Woody. The family desired to see the company continue as a family-owned, family-operated business long into the future. The family's youngsters are offered employment in the company, but by no means is there a guarantee of an executive role for a family member. However, if one of the family members falls upon hard times, the remainder of the family and the resources of the company are there to help them. As Howard put it, "What good is it owning a company if I can't offer jobs to my family when they need them?"

If one went into the Union offices in Mountain View during the centennial year, she or he would find a beehive of activity. The entryway serves as a store for the latest in personal communications technology. The staff in this room handles everything from providing customer services over the phone to taking the payments from customers who wish to save postage or prefer to pay with cash. In an industry where companies often lived and died by their service contracts, Union tried something a little different. They offered prospective customers who were locked into a contract with a competitor the option of having Union pay off that contract in return for the consumer signing on to a Union Wireless agreement. The strategy has paid off, and it has brought in many new customers. Competitors countered with plans that lack annual service contracts.

Behind the retail-store room in the entryway is the main office complex. Here, one sees a poster with the Ten Principles of Cowboy Ethics, written by James P. Owen, a Wall Street businessman and author of a best-selling book about how to pick up women. The principles are intended to make business philosophies sound kind of sage-brushy. However, several of the main beliefs in the list are important for businesses, especially a company like Union Wireless. One gets the impression that the Union employees really do as Owen suggests: they "ride for the brand."

Standing, from left to right: John G. Woody (third generation), Brian Woody (fourth generation), Eric Woody (fourth generation), and Stacey Aughe (fourth generation). Seated: Howard D. Woody (second generation).

There are three members of generation four serving on the board: Eric Woody, Stacey Aughe, and Robert Brian Woody (called Brian). Of the ten members of this generation, three chose not to become board members. James D. Woody had worked hard for the company and was on track for a leadership role when a sudden illness claimed his young life. Other members of the family chose not to take the path towards leading the board. For example, Howard G. Woody, called Gene, works for the company, but—by his own account—he was not willing to take on the responsibilities and put in all the additional time that being a board member requires. However, he remains a valuable part of the company. No one in the family is forced to participate in the company more than they want to.

MS. INFORMATION

Operators get some of the strangest calls. The following is an example of an actual call to the Union switchboard.

Operator: "Information—how can I help you?"

Caller: "Could you tell me the number for 9-1-1?"

All of the members of generation four expressed similar recollections about not being prodded to seek a leadership position in the company. This generation was the first to go through the local public school system after the company became a major employer in the local community. A few have stories about being treated differently by their classmates, but for the most part they had typical childhoods in a rural community. Apparently, many people in the local community thought the Woody family members all had access to the best and most up-to-date cell phones and communications devices and were enormously wealthy. But the company does not hand out mobile phones for the children in the family, and none of the generation-four members felt like they were born with sliver spoons in their mouths. They are all well-grounded, levelheaded individuals who do not put on any airs of superiority.

Howard's "boot camp" at his 7,400-acre ranch kept the youngsters free of illusions of grandeur, and it prepared one and all for a life of hard work and responsibility. The generation-four members recalled helping Howard to hay, to brand, or to do whatever else was needed of them to take care of the ranch. In many respects, the upbringing of generation four was much like that of generations two and three. Once again, the ranch served as a vehicle to instill dependability and a respect for traditions. The ranch has always played a vital role in teaching the younger generations the need for teamwork, hard work, and dedication. It is where they learn, as Howard says: "There's one thing I always taught'em: the company comes first. And that is the truth. If you take care of the company, it will always take care of you. It was something I tried to instill in all the children."

Haying and branding times in Mountain View are events that bring generations together to work the land and touch the spirits of Azariah, John D., and the others who now walk in the shadowlands. Now in his nineties, Howard *could* afford to have the hay brought in or have workers cut it for him. However, purchasing all the hay required to feed the family-owned livestock might throw away the opportunity to teach the coming generations a little about life lessons and the family traditions. There are still some things that money can't buy.

All of the generation-four members sitting on the board of directors have college degrees and have worked outside of the family business. For example, Eric served as a nuclear technician in the US Navy's submarine service. After the military, Eric obtained an electrical engineering degree. Brian worked as a programmer and a machinist before and during his pursuit of an electrical engineering degree. The generation-four members feel this kind of outside work allowed them to bring other experiences into the company.

Stacey also served in the military. Her army career took her to far-flung places such as Korea, Fort Burgwin, and Walter Reed Army Medical Center. She specialized in radio repair, but also cross-trained in satellite-systems repair and audio/visual systems. She is a member of the American Legion. After her time in the army, Stacey earned a master's degree in electrical engineering from Clarkson University in Potsdam, New York. When asked where she acquired her interest in electronics, she jokes that she learned by osmosis. She recalled spending many hours at Union Telephone in the company of her mother, Gloria Dee Woody. "I was actually a nosey child. So I always wanted to know how things worked." Having been raised in an environment populated by engineers and furnished with advanced telecommunications equipment, it is little wonder that the inquisitive young girl hanging around the office became one of the company's leading engineers and a member of the board of directors.

Of the generation-four members, Eric and Brian seem the closest. They worked together for years sharing a truck and a cubicle. In the centennial year, they still shared an office. They trade humorous barbs with one another across their equally cluttered desks, but always with great respect. They recounted a period when they weren't talking with one another and communicated in grunts and by throwing "stress balls" at each other. One of them remembers the "split" being a few weeks long, and the other remembers a period of roughly six months. This difference in the recollection was enough to set off another good-natured debate. Neither could remember the cause of the dispute, but both remember the fun they had with each other in expressing their mutual displeasure. Eric and Brian evolved into partners, even to the point that they were always recommending each other for important activities and projects.

Making the most of their college educations, generation four is the most technologically savvy generation of Union's first hundred years, and they prepared long and hard for their current roles. Additionally, the generation continues the tradition of having hands-on experience with the technology and the rigors of constructing and maintaining the systems in the field. They're clearly proud of their understanding of how the systems work and what it takes to keep complex electrical systems functioning in America's Outback.

The generation-four members thrive on tackling problems and coming up with innovative solutions. It seems to be more than just an "engineer thing." For example, Brian has a passion for figuring out devices that could help those with hearing disabilities. Since Brian's uncle John G. is hard of hearing, there was a handy guinea pig to test the latest equipment configurations upon. One might think that running the company must keep a person busy, but there was Brian, taking the time to link a Bluetooth device to his uncle's hearing aids as if it was his only job.

Generation four was the first generation to work its way through the succession system that was so carefully developed by the family. The members of the generation were active in shaping the succession plan with an eye clearly toward improving the competitiveness of the company. Generation-four members were deeply involved in crafting the 2009 modifications to the company's bylaws that effectively ended the management team approach and replaced it with a corporate-executive structure. Generation four is a group of dedicated go-getters. The "G4s," as they are sometimes called, share an easy sense of humor and self-effacing style of dealing with outsiders. They compete with one another to a certain extent, but it is abundantly clear that they all have each other's backs and respect the talents of each member.

FILLING IN THE MAP

Wyoming is a very wide town connected by very long streets.
—Eric Woody, recalling a conversation with Mrs. Fields of Newcastle

In the 1990s, Union Telephone continued to add exchanges to its territory. In 1994, the Saratoga, Encampment, Shirley Basin, La Barge, Hanna, and Rock River exchanges were added to the Union system. It was not an easy transition to bring old, established exchanges into a new network. The process can take years, and it was far more complex than in previous times. Extended public processes and compliance with regulations replaced simple business deals. In the case of the Saratoga and Encampment exchanges, there initially was a significant local concern about losing their old services. The company that sold the exchanges to Union Telephone could have made the transition smoother, but the Union staff worked assiduously to allay the concerns of their new customers. With the passage of time, the customers adjusted to the changes, and many of them transferred their "home team" loyalties to Union.

Union continued to expand its cellular coverage as well. In May of 1994, there were roughly six thousand cellular customers, a figure that more than quintupled to thirty-one thousand by December of 1998. During this process, Union was able to extend coverage within their territory by adding more cellular towers. For example, in the early 2000s, Union constructed a cellular tower near the Gates of Lodore, which expanded service into the lower end of Brown's Park. By Union's centennial year, the company had built towers in Montana, Utah, Colorado, Idaho, and Wyoming. The company had grown significantly such that larger companies with more resources started paying attention to Union. Eric Woody compares Union's growth model with their competitors: "They look for where people are, and we look where people go through."

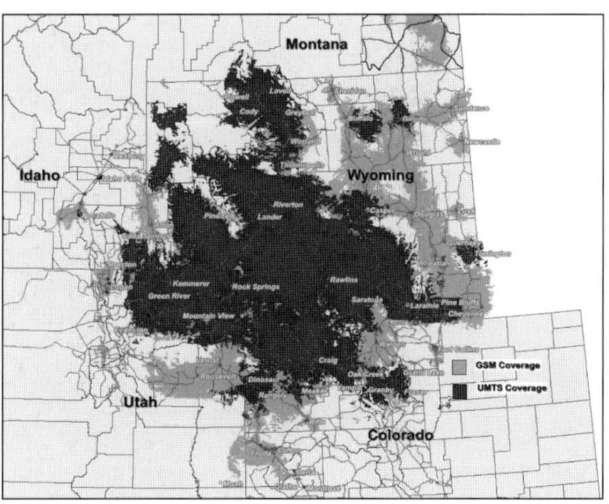

The Union Telephone service map in its centennial year.

Part of the company's growth in the 1990s was acquisition of the Saratoga exchange.

With the smallest population of any state, Wyoming remains a somewhat unattractive prospect to the boards of other corporations. Competitors can see the returns on investments that Union has earned, but the length of time necessary to achieve such returns is far too long for many competitors to jump into the outback. As Union Wireless continues to improve its competitive position, grappling with the tentacles of other telecommunications companies is a distinct possibility.

RETAIL STORES

In 1990, Union Telephone began opening retail stores in order to supplement their Mountain View store. The stores have showrooms where potential customers can try out the communications devices, buy accessories, discuss service plans, and even pay their telephone bill. In 1994, the company added stores in Saratoga, Steamboat, and Jackson. As the retail stores proved successful, more of them were added to the system in towns such as Rock Springs, Casper, and Rawlins. Since that time, stores continue to spring up on the map. The fifteenth retail store is scheduled to open during Union's centennial year.

The retail stores help give Union Telephone a physical presence in many of the communities it serves. The stores are instrumental in promoting the growth of the Union Wireless customer base. Each store employs roughly one supervisor and three retail sales associates. The growth in the retail side of the business has increased the Union Telephone workforce. As the company earned a reputation as a good employer, it became easier to attract better employees to the company.

In its centennial year, Union operated fifteen retail stores. This store is located in Rock Spring, Wyoming.

UNION HOLDING CORPORATION

In 2010, Union formed a holding corporation for the telephone company's stock. This step was taken at the behest of CoBank in order to secure a mortgage. For tax purposes, this company was incorporated under the laws of Delaware. The Delaware company contains the stocks held by the original holders of Union Telephone Company. This allows all the stock to be kept to secure Union's loans.

The holding corporation has further advantages, as the last of the remaining minority shareholders retire their stock and the family is able to draw upon the value of the stock. Additionally, the Delaware company provides a mechanism for the family to hold other companies that might be purchased in the future.

BEING THE GOOD NEIGHBOR

We are pleased to partner with the Wyoming Food Bank of the Rockies again to help end hunger in Wyoming. Involvement with organizations such as the Food Bank is just one of the many ways that Union invests in the community.
—Brian Woody in a press release

Union Telephone Company has a long track record of living up to its good-neighbor image. The company established a charitable foundation, the Linda Kay Woody Memorial foundation, to provide childcare for those who cannot afford it. Under Eric Woody's guidance, Union Telephone has provided scholarships for local high schools, the University of Wyoming, and Utah State University. Both Brian and Eric were actively engaged with the Boy Scouts of America (BSA) as scoutmasters. For many years, Howard worked with the BSA as well. Eric has worked with the Competitive Carriers Association's "CCA Gives Back" program.

MS. INFORMATION

Operators get some of the strangest calls. The following is an example of an actual call to the Union switchboard.

Operator: "Union Telephone Operator—how can I help you?"

Caller: "Is that anything like Roto-Rooter?"

Brian and Eric have also worked with the Special Olympics. The company has provided funding for food banks, and it is a generous presence in the local community, evidenced by their sponsorship of teams and events.

Due to the increasing amount of wildfires (which have been prevalent in the Rocky Mountain West since the 1990s), the company put together a cellular on wheels (COW) system of mobile towers. This ingenious system allows firefighters and support personnel to stay in communication in areas that usually do not have cellular service. This system produces more-secure field communications and allows the firefighters to keep in touch with loved ones while away from home.

BOOMS AND BUSTS

In the 1990s (and extending into the new millennium), Wyoming had yet another energy boom. This economic expansion was spurred by natural-gas extraction. New technology made it possible to extract gas from the tight sands of vast stretches of Wyoming's intermountain basins. Tight-sands tax incentives made the development of new technologies and field infrastructure more attractive. Places like Jonah, Wamsutter, Rawlins, Pinedale, Opal, Lyman, Big Piney, Granger, Casper, and many other places once again became hubs of rapid industrial development. This boom spread to northern Colorado as well. Across the region, thousands of gas wells were drilled each year. Huge pipeline systems were put in the ground to take the product to market. Compressor stations (essentially complex banks of high-powered fans) provided power to move the gas by pressurizing it.

Much of the gas from the region is the highly poisonous "sour gas" that has to be treated by removing sulfur dioxide. The flares of gas that emanate from the large industrial complexes used to "sweeten" the gas can be seen for miles in the Wyoming night skies. Large crews of "thumper" vehicles ply the backcountry, sending vibrations through the ground that are read by sophisticated computers to predict where the best pockets of gas are located.

The boom in the 1990s made the 1970s boom seem like a quiet time. The nineties ushered in a sustained boom of several years that fluctuated with the spot prices of the gas market. Workers came into the expanding Union Telephone service areas from all over the world. There were seismic explorations crews from Thailand and Mexico, drillers from Canada, engineers from Britain, as well as plenty of Texans and Oklahomans. Large gas fields were developed by foreign corporations such as EnCana (Canadian), Shell (Dutch), and British Petroleum.

A main challenge faced by generation four has been to position itself to be responsive to the growing and changing communications needs of the energy industries. This is an extension of the model that generations two and three used to help build the company and steer it toward profitability. The energy companies have deep pockets and need reliable communications wherever they next choose to build.

The companies associated with energy extraction are important customers, but they are sometimes difficult ones to keep satisfied. Like the products they produce, the gas industry is volatile. A gas field near Wamsutter, Wyoming, may be popular for a few years, only to have the focus of activity move miles away to Hiawatha with a new hot spot maybe only a few months in the future. Promising results from this year's seismic exploration might generate a rush of exploratory wells next year in a place previously believed to have little potential for development.

Union's communications network has to nimbly adapt to the changing requirements of the industry, and management has made shrewd decisions on where to place their towers and other infrastructure in order to keep up. Union's management pays attention to where the industry's next potential gas fields are likely to spring up, and they plan their tower-construction program accordingly. As always, their customers are quick to request services when Union does not have them pre-positioned.

During this time, Union Wireless was exceptionally well placed to pick up much of the "gas patch" business. Cellular service in the developing fields became a necessity; it was no longer simply the luxury it once had been. Union Telephone prided itself on its coverage in the main fields such as the Jonah field, the Anticline, and Moxa Arch. The intermittent busts resulted in downturns in service needs, but Union Wireless continued to fill in their system, recognizing that boom/bust events are cyclical and that the time to prepare for the next boom is during the bust.

GATED MODE ACQUISITION (GMA)

In 2007, Union learned of an initiative regarding open connectivity, which is a roaming hubbing model in the Gated Mode Acquisition (GMA) world. Union saw the model as a means to reduce the overall costs related to establishing roaming relationships with mainly global partners. Unlike the more familiar unilateral agreements, this model employs third-party partners in a bilateral agreement in which a service company arranges to pay for access to the system of another provider.

Union began establishing bilateral roaming agreements with foreign service providers. This allowed Union customers to travel in foreign nations where there is a bilateral agreement without having to change their subscriber identity module (SIM) card or change their telephone number. Similarly, foreign visitors coming to Yellowstone or working in Wyoming's minerals-extraction industries would be permitted to retain their local carrier and use Union's system of services. The bilateral agreements opened up the potential for new revenue streams for Union. However, the ability to perform this magic requires that the systems involved in these open-connectivity arrangements have the same technical standards.

Eric participated in the development of standards for open connectivity. Without the participation of smaller service providers like Union, there was a possibility that the standards would benefit the telecommunications giants and effectively shut out the small businesses. Union volunteered to be one of the trial partners for testing the standards. This test put Union in a relationship with a company in Mongolia, which Eric said "created its own entertainment." The test linked Union through a roaming hub in the United States, then through a roaming hub in Europe, and finally to the Mongolian company. Eric stated: "It was really complicated, but it gave us a taste for what was possible. At the end of the trial we found out it worked."

The test had shown that the then-current standards (which relied upon cumbersome network extensions) could be improved. The network extensions required individuals wishing to internationally roam to have multiple profiles for their phone's SIM (subscriber identity module) profile, which made the phone appear to be a subscriber of the company with which Union had a unilateral roaming agreement. Roaming hubs made it possible for Union customers to roam in areas for which there were roaming agreements without having to alter their SIM profile or prepay for the service.

Even with successful tests, roaming hubs were not a sure thing. Union participated in the meeting in Germany that passed—by a very narrow margin—this measure as a business standard. The trial Union participated in did not indicate what it would cost to run such a service commercially, but Union had helped pioneer that it was technologically feasible. Eric laughed as he discussed the concept of Union as a pioneer, saying, "It kind of fits our mantra."

When asked if participation in so many boards, groups, and associations created a burden upon the company, both Brian and Eric indicated that there was somewhat of a logistical burden in participating in so many activities scattered across the globe. Brian remarked, "However, you get into a space where you can steer the associations into really helping your company." Having a voice at the table has proven to be important.

UNION INTERNATIONAL

The jump to cellular service kicked open the door of possibility for the telephone to become a preferred entryway to the World Wide Web/Internet. The early phones were somewhat slow. Downloading content and inputting alphabetic characters was challenging. More-recent versions of the technology truly make the telephone a portal to the wider world. Union customers found themselves not only able to surf the Internet to find the best restaurants in Portugal, but also to dine at one of those restaurants and simultaneously upload audio and video content of the experience for faraway friends and family. The customers continued to inform Union as to what services they desired, and they demanded international-service bargains. By the same token, foreign customers preferred that their phone provider serve them on their business trip or vacation in the United States. In order to retain customers, telephone-service providers increasingly had to keep their eyes on the international-service markets.

By its centennial year, Union Wireless was aggressive in its pursuit of agreements with international providers, adding an international capability to the Union Extended Wireless network. European customers began roaming Union's systems in 2006. During the first month the service was available, the company saw revenue of roughly $30,000 from that source. However, it took substantial effort to get Union customers to sign agreements for roaming services outside North America. In 2009, Union first offered European roaming services to its customers.

Eric Woody, the chief technology and operations officer, spent much of his time traveling overseas to negotiate these agreements. In its centennial year, the company had over 150 roaming agreements outside of North America, and it was looking to expand into the largest of global markets: China. Perhaps "Union Global" will be a new iteration of the company in the coming decades.

There was a learning curve when Union entered the international marketplace. As a local and regional provider, Union clearly understood the customs and concerns of its clientele, its competitors, and its partners. According to Eric: "We had difficulties getting international-roaming agreements, mainly for the same reason people do business here. If they don't know who they are doing business with, no matter what the opportunity, they're leery to enter into it. We found that Europe especially is relationship based. They don't want to do business with a company they don't know."

Eric packed his luggage and began to establish both a corporate and a personal face for the company. He first went to London for a meeting to finalize the standards for open connectivity, and he began the long process of introductions and meetings. He learned that in foreign markets, the reasons for doing business are every bit as important as making money.

Union took advantage of opportunities that arose to gain toeholds in the global marketplace. A successful strategy for Union had been to participate in telecommunications-industry working groups, committees, boards, and associations. Eric attended the lion's share of the international corporate meetings, but Brian participated in this capacity as well. Both Eric and Brian worked the numerous trade shows and industry meetings, such as the GSMA's World Congress and the CES (Consumer Electronics Show).

When asked which industry associations the two participate in, the response was an alphabet soup of acronyms mostly associated with the GSMA (Groupe Spéciale Mobile Association). A list of the groups and associations includes, but is not limited to: CCA (Competitive Carriers Association); BARG (Billing Accounting Roaming Group); TADIG (Transfer Accounts Data Interexchange

Group); RING (Roaming Innovation Group); TWG (Terminals Working Group); RCPG (Roaming Charging Principles Group); FSWG (Fraud and Security Working Group); FSCIG (Financial Settlement Clearing Interest Group); North America Group; SSCG (Standards and Smart Card Group); CSRIC (Communications Security Reliability and Interoperability Council); RCS Task Force; RWA (Rural Wireless Association); and NTCA (National Telecommunications Cooperative Association). All of the groups within the GSMA are in the process of being renamed. For example, BARG is becoming BG (Business Group). And the beat goes on.

In addition to activities within such associations and groups, Union representatives participate in business meetings across the world, such as GSMA meetings in Latin America and Asia. Brian and Eric each make roughly four trips per year to Washington, DC, to lobby members of the US Congress and visit the Washington offices of federal regulatory agencies. Unlike many lobbyists, they do not go to Congress with their checkbooks out. They seek to make sure their representatives know where Union stands on issues affecting telecommunications, and they use the advantages associated with coming from a state where people tend to know one another.

It helps Union that the gas fields in Wyoming and northern Colorado are the busiest in North America, and that multinational corporations are developing much of the natural gas from the area. Many of the individuals and compa-

MS. INFORMATION

Operators get some of the strangest calls. The following is an example of an actual call to the Union switchboard. This caller must have thought that operators are all knowing and keep track of every new business in every town.

Operator: "Directory assistance—how can I help you?"

Caller: "I want a number for [garbled]."

Operator: "Where?"

Caller: "Just up the hill in Urie. That new place on the corner."

nies working in "gas patch" are from Canada, Brazil, and Australia, and these roaming partnerships help keep the business community—so important to the local economy—connected.

International agreements opened revenue streams previously unavailable to the company, and they provided customers with the services they requested. Growth in international roaming is the trend. Eric provided the following information regarding international roaming. "In 2008, our total international [revenue] was under $2,000. [In] 2009, [it was] under $100,000, 2010 it was $250,000, 2011 it was $220,000, 2012 it was just under $400,000, [and] 2013 it was about $350,000." The trends also evidence increased data usage and text messaging.

Eric seemed especially fond of setting up the roaming agreements in the Bahamas. Perhaps a portion of his passion for establishing such agreements relates to his inner adventurer that always seeks new horizons. The future had knocked on the door, and Union had answered.

THERE IS NO SUCH THING AS BAD PUBLICITY?

It is better to ask forgiveness than seek permission.
—*Old English proverb*

In 2009, a homeowner near Alcova, Wyoming, noticed a cell tower near his home, and he contacted the Natrona County planning commission to find out how the tower had been approved. The inquiry set off by this homeowner tarnished Union's good-neighbor image. While the planning commission had approved a tower nearby, the as-built locations for this tower were a mile from the permitted location. The commissioners could not understand why the tower was not in the approved place.

Union Telephone's spokesperson for the incident, Eric Woody, originally suggested that it might have been a mistake, since this particular tower's proposed location had been changed six times before it was constructed. Confusion in the planning process is not a new story in the complex land-ownership pattern of Wyoming. However, as Eric Woody looked into the situation, it became clear that confusion was not the issue that had caused this tower's erroneous placement. Union Telephone had applied for a permit for the equipment building located on the compound, but they had not applied for a permit for a tower to be built on the site. Something had gone terribly wrong. Eric stated: "Now it's a matter of finding out where we're at. Basically finding out how big a mess we have." The mess included several zoning infractions.

The ensuing tempest of media reports about Union Telephone was less than flattering. The *Casper Star-Tribune* ran articles with clever headings like "Phone Company Puts County on Hold." Suddenly, Union was in the uncomfortable position of being in the public spotlight, as agencies investigated other Union Wireless towers to see if they were in compliance. There was the expected grandstanding and chest pounding in the press by local politicians.

The errant tower created a crucible within which the generation-four members demonstrated what they could do. One can learn a lot about riding in that moment of taking a spill. Some feel that moment is worth many hours of successfully staying in the saddle. But no one will argue that taking a spill is desirable.

According to Eric Woody, an employee who acted outside of the company's policies apparently made the decision to move the tower from its approved location. Union Wireless management did not approve this action, and they were just as surprised as the commissioners when the problem came to light. The individual responsible for moving the tower was fired, and the planning commission assessed Union an "aggravation fine" (also called an "impact fee" in the language of bureaucracy) of $10,000.

As a result of this incident, Union Telephone added more checks and balances to its internal system to ensure that future construction projects corresponded with the approved sites. The wisest reaction to this incident occurred when Union representatives went back to Natrona County and offered to do whatever the county demanded to remedy the issue. That response was a good reminder that Union Telephone was not an outsider and wished to remain a good neighbor. Outsiders would have shown up with teams of lawyers and responded with high-power political pressure against the regulators. Union clearly wished to retain the good will of their neighbors in Natrona County, and they were more than willing to set things right. The company was in Natrona County for the long run, and it wished to restore the trust of the county and its inhabitants.

There were continuing indications in the press that there may have been as many as five towers built without all of the appropriate approvals from Natrona County. That appeared to have been some of the hyperbole of the time, as no similar wrongdoing was brought to light.

The required permit applications were filed and eventually approved for the errant tower. Even then, the press did not cool down for some time. The *Casper Star-Tribune* announced a settlement of the issue with the contradictory heading, "Planning Commission OKs Illegal Tower."

This turned out to be a test for generation four. They became involved in handling the highly charged meetings with the planning commission and dealing with many of the press contacts related to the incident. Eric had only recently been promoted to the board of directors, and he had nightmares over dealing with the situation before it was over. He and the rest of the management team brought the company through these stormy seas, and there has never been a repeat of the situation. Union Telephone returned to wearing its good-neighbor hat, much wiser from the experience of almost losing that status.

THIS PHONE IS SO FAST, IT GETS CALLS FROM THE FUTURE

The communications industry in general has been placed on a track that truly gives the race to the swiftest. Regarding technological changes, Union Telephone and its competitors each strive to keep one step ahead of one another. Fast 2G systems were rapidly outpaced and replaced by 3G systems. These, in turn, were left in the dust of 4G systems. The *G* in all of these systems stands for generation. The switch from 1G to 2G was a switch from analog to digital systems. The jump to 3G provided mobile broadband, and 4G is advertised as allowing faster data transfer, though some 3G systems were just as fast. The bottom line for all of these generations of technology is that service providers need to go through expensive upgrades with each new generation. The cost for Union Telephone to upgrade from 3G to 4G was roughly thirteen million dollars. On top of this, providers must make telephones that use the new technology available to consumers, and subsequently convince them that there are advantages to buying in to the next generation. As this book was being researched, there already was talk about upgrading to 5G. The best is yet to come.

If the present is an accurate predictor of the future, companies like Union Telephone are going to be in a constant cycle of upgrading, marketing the new technology, and putting aside capital for the next upgrade. If technological generations are roughly five years apart, then this dovetails well with a five-year corporate plan. However, what if generation upgrades start coming along only one or two years apart? With the successful branding of smart phones, companies have shown that the customers seem willing to switch to the newest model, even when there is roughly a year between iterations of the devices.

A typical Union tower, this one located near Point of Rocks, Wyoming.

YOU STILL WALK THE FERTILE FIELDS OF MY MIND

Warm summer sun,
    shine brightly here,
Warm Southern wind,
    blow softly here,
Green sod above,
    lie light, lie light,
Good night, dear heart;
    good night, good night.

—Mark Twain's 1896 eulogy for his daughter Olivia

Linda K. Woody

The loss of loved ones comes to us all. In the period of a few years, the Woody family endured the passing of several members. It started with the loss of John G. Woody's wife, Linda Kay, in 1997. As a memorial, the family set up a charity foundation in her name that helps children.

Esther and Dee soon followed Linda in 2004. The sudden deaths of these two key members of the family in such a short period of time rocked the family and the business. Howard had lost the love of his life, Esther, and his beloved daughter within a few months of one another. He was despondent for years, but did his best to hide it. One can still see him fighting back the pain when he recalled these times. Many people did not expect him to get through this period, and there was a natural tendency to try to shelter and assist him at this time. Neighbors often checked in on Howard to see how he was doing.

Similarly, the generation-three members lost both their mother and the sister who was so very much like their mother. Generation-four members had lost a mother, a grandmother, and an aunt. Their families and loved ones comforted them during these times. Each person dealt with the losses in their own manner.

James D. Woody

The Woody family showed remarkable resilience in these troubled times. Rather than withdraw into a shell of grief, they showed up at work, like always, and continued on seemingly as if nothing had happened. But something profound did happen, and each member went through the grieving process at their own pace and in their own fashion. Internalizing such strong emotions might not be the healthiest way to deal with such profound losses, but it fit with the family. Both Esther and Dee were key personalities who had influenced many of the company's operations. Esther had been the glue that held the family together, and Dee had been a driving force on the board of directors. They didn't leave an empty space—they left a black hole.

As if this was not enough, James H. Woody's son James D. passed away in 2005. Then Lottiebelle Woody followed in 2006. Lee J. Woody, the quiet and calm voice on the board, joined them in 2007. Once again, the family outwardly appeared to shelve their grief and "cowboyed" on. It was their way, and it was only by drawing on those family ties that they found the personal strength to endure, as William Shakespeare refers to them, "the slings and arrows of outrageous fortune."

TIME IS ON MY SIDE

In Union's centennial year, three members of generation four were serving on the board. Two members of generation three have passed on, two still serve on the board, and one did not become a board member. The two members of generation three on the board are "cutting back" their involvement, and they're positioned to "retire," if any Woody family member knows how to do that. The strong family work ethic does not allow a Woody to gracefully head for the rocking chair on the porch to spend the remainder of his or her days whittling. That would be out of character. Although retired, Howard still oversees everything outside of the day-to-day operations, and he remains the guiding light of the business. The generation-four members have worked hard and paid their dues to serve on the board of directors. They have more votes than any generation, but they're still considered the junior members, even though they're really running the show. This will inevitably change, and incoming generation-five members will become the new juniors.

The previous generations sing the praises of their descendants in generation four, recognizing great growth and untapped potential in them. Generations two and three are very proud of the generation-four members. Generation-three member Jim Woody summed it up: "They have a little different attitude. They don't remember it being a five-man outfit or even a twenty-man outfit. They have missed a lot of the background that my generation had. At the same time, they know a lot of things that we don't know . . . I don't see them not going forward in the business in an appropriate manner. Although, they still have to check in with us old guys."

After more than a decade of working with the consultants on communication issues, the investment has paid off, and the torch has successfully been handed to the new generation. "Gen-4" has hit the ground running, and its members have genuinely bought into the program. Brian explains: "It's not a job. This is what our life is. Yeah, we have pastimes when we go home to our families. We would like them to participate in this as much as they are willing. At the end of the day, I hope I have the drive that Grandfather has."

Our next connection takes us to the end of our century-long journey, and it delivers a message from the past generations to those who one day will steer the corporate ship through Union Telephone's second century.

Chapter 6

The Fifth Generation: Keeping the Dream Alive

Trying to predict the future is like trying to drive down a country road at night with no lights while looking out the back window.

—Peter F. Drucker

QUO VADIS?[1]

The forthcoming pages in this tale of connections are for the Woody family descendants, the employees, and the customers to write. The company started out as the vision of the hard-working, soft-spoken dreamer John D. Woody. His descendants have taken Union Telephone Company from a single employee who was supported by unpaid family helpers to a corporation that employs hundreds. The company has grown from a service that connected a few isolated ranch telephone lines to a powerhouse in communications, with infrastructure serving large parts of the Rocky Mountain West and customers numbering tens of thousands. Though the roots of the company remain firmly planted in Wyoming soil, the branches of this tree have spread service across the nation, and at the time of the company's centennial, Union was making inroads on the global-services market.

In the centennial year, the fifth generation was in the early stages of the process of earning their respective places on the board. Those raised in the Mountain View area were often the butt of jokes about the publicly perceived privileges of being from one of the better employers in the Bridger Valley.

[1] *For those who don't read Latin, this translates as "Where ya headed?" in Wyoming-English.*

This photograph shows Howard's boot camp.

According to generation-five member Michael Woody: "You know people always leave their little two cents. . . . Really, that doesn't bother me because the company was made to provide jobs for the family. . . . Union's really a great company. I'm not ashamed to be part of that."

As children, the generation-five members all worked at the ranch with Howard. This remains the tradition for inculcating family members with the company's work ethic. Those that wish to rise to a leadership position within the company will need to become much like their elders. They will need to learn about the business via immersion in a hands-on approach. They will need to undertake useful education and be able to apply it to the practical issues of building and maintaining whatever forms the technology evolves into. They will learn about the deepest corners of the American Outback, and master the challenges that nature presents to those who would seek to flourish there. They will find the spirits of their ancestors calling to them from windswept, frozen peaks. In such places, they will discover what made the previous generations special, and they will come to admire the amazing things that the family has accomplished.

Some of the generation-five members have already chosen to take employment in the company. They learned the ropes by building towers and working their way through various departments. Abbey Woody held jobs outside of Union

MS. INFORMATION

Operators get some of the strangest calls. The following is an example of an actual call to the Union switchboard.

One day a customer called with a complaint about her phone service. The operator asked, "Who is your phone provider?" Without hesitation, the caller replied, "My mom."

Telephone before she decided: "It was time to go back to the family business. It was something I grew up with and knew about." The generation-five members who were interviewed expressed a desire to keep the company on the current course. They have no secret plan to change the corporate structure or culture. But then, generations two through four all might have similarly answered the question of where they planned to lead the company. The previous generations did not envision radiotelephones, microwave transmitters, digital switches, cellular systems, smart phones, and the wireless world of the present. All of these technologies forced fundamental changes in how the business provided services. Change will always be there. As long as each generation's members can abide by the family traditions of hard work and service to neighbors, the technology is just window dressing wrapped around the original dream of the company's founder.

What has been instrumental to Union Telephone's survival is the unwavering commitment to make John D. Woody's dream of connecting neighbors a reality. Combined with a strong work ethic and a family that raises members who are strong enough to disagree with each other but levelheaded enough to put aside rivalries for the good of the family, the generation-five members started out with tremendous advantages. They are technologically savvy, and they are committed to keeping the company on the cutting edge. The generation-five members see their parents and Howard as their role models. Behind them all, John D. Woody's spirit remains the beacon that has shown the family its path through the company's first century. That is a good start for getting to a bicentennial celebration in 2114.

The ten members of generation five have only begun their careers with the company. The realization that one or more of them will be running the company in the future is only starting to sink in. They express their belief that the company is being run well, and they don't foresee that there will be major changes to make. As Michael Woody put it, "I believe if we continue to run the company the way that we are, it will continue to prosper." Abbey Woody concurred, and added that the company's model of treating the customers like neighbors and not just numbers would serve it well into the future, setting Union apart from its rivals.

SEMI-RETIRED, AND LOVING IT

One thing that is clear about the family members that have served on the board and the management team is that they may retire, but they continue to work. In the centennial year, Howard, John G., and Jim all claimed to be "semi-retired" or less active in daily management. Semi-retirement is an idyllic corporate state in which an individual gets to hand off the responsibilities of dealing with the day-to-day operations of the company to the next generation while still showing up to make sure that the new managers keep the course straight and true. This is similar to hanging on to the remote control for an appliance that is being used by someone else. The semi-retirement path allows the previous and outgoing leadership to let go, but still remain involved. It is not the easiest way for new managers to learn the ropes of running the business, but it maintains continuity.

For the semi-retired, it can be the best of both worlds. Those semi-retired folks do not put in as many regular hours as they did before entering into this sublime state of existence. The next generation gets to do the "grunt work," while the previous generations remind them of what the company's struggles were like in the past. Most people who retire leave the job behind and move on to new things. But semi-retired folks essentially become both mentors and backseat drivers. Who would not want that deal?

Given that the Woody family still raises their children in a manner similar to Howard's upbringing, one has to believe that generation four will semi-retire in about two decades. That means that like generations three and four, generation five will receive the benefit of on-tap corporate memory. If they don't mind having to look back over their shoulders as they deal with an industry where change is exponential, things will work out.

ARE THE FIRST HUNDRED YEARS THE EASIEST OR THE TOUGHEST?

Every calamity is to be overcome by endurance.
—*Virgil*

Will the coming generations be able to ensure that the company survives to a bicentennial celebration? The odds are against it, but then if the Woody family had played the odds, there would have been no centennial celebration for Union Telephone. If Mountain Bell/AT&T had taken that offer decades ago to buy the company for a dollar, Union would have gone the way of most of its competitors. Some might say destiny had a role in bringing the company to its centennial year. The leadership of the company has spent significant time, capital, and effort towards guaranteeing that the following generations have stability, ability, and the resources necessary to compete.

The company acquired bandwidth that is currently not in use on the assumption that someday there will be technology that uses that precious commodity. The bandwidth is a system of intangible lines waiting for the day when they are needed. This is much like a rancher picking up a parcel of scrubland on the bet that someday the acreage will be usable. It is a strategy that has served the company well on its trip through time. Small exchanges that were picked up when they were pretty much worthless later became valuable when development by extractive industries put the system in the middle of a series of economic booms. Gas, oil, coal, and trona all created demand for telecommunications in the Union Telephone service areas. The discovery of lithium near Point-of-Rocks, Wyoming, may generate yet another such uptick, one for which Union is superbly placed to provide necessary telecommunications services.

In a field as rapidly changing as telecommunications, the path forward is difficult to predict. Futurists have been trying to foretell the future for Mr. Bell's wondrous invention ever since it passed from novelty to practicality. People in the industry seem to declare that communication devices will get smaller, faster, and smarter. Watching an older member of the Woody family receive a telephone call through his hearing aides is a clear indication that the engineers are just beginning to plumb the depths of where telecommunications can go. Communications devices implanted in prosthetics are within current technological capabilities. Contact lenses that monitor the blood sugar of diabetics and allow the user to download that information have become reality, not science fiction. Perhaps someday a person will be notified to see a doctor when one of the monitoring/communications devices implanted in their body alerts them that a visit is required. These devices could mark the end of annual medical checkups. No doubt, future devices could possibly dispense various medications upon receiving an electronic prescription from a doctor.

Will the trend in miniaturization lead to communications gadgets being part of the fillings in our teeth? Will our communications devices eventually become our alter egos, taking on our personalities or even developing their own? The thought of a personal device nagging us to behave in certain ways is unsettling. Yet, such contraptions already exist. Will our communications devices replace all cash, credit cards, and identification cards? Will they replace pets as companions? Will the new inventions translate our thoughts into any language in order to break down those ancient communications barriers that keep people apart? If current trends continue, what will happen to individual privacy? Will the communications instruments of the future become so indispensable that it will be impossible to function in society without at least one? Will the devices stifle political change or create a bold new experience in democracy through electronic voting?

Whatever future innovations are made in technology, generation five is going to have to incorporate them into their version of Union Telephone. They will need to acquire the skills to rapidly and correctly respond to these changes. Change is coming like a speeding freight train. The family is doing its best to make sure they will be ready for whatever comes their way.

Writing in 1910, Herbert N. Casson wrote about his perceived future for the phone.

> Some day—who knows?—there may come the poetry and grand opera of the telephone. Artists may come who will portray the marvel of the wires that quiver with electrified words, and the romance of the switchboards that tremble with the secrets of a great city. Already Puvis de Chavannes, by one of his superb panels in the Boston Library, has admitted the telephone and the telegraph to the world of art. He has embodied them as two flying figures, poised above the electric wires, and with the following inscription underneath: "By the wondrous agency of electricity, speech dashes through space and swift as lightning bears tidings of good and evil." (Casson 1910, 72)

Perhaps it is most appropriate in this review of the past hundred years—which took telecommunications from the fences in the fields to satellites orbiting in space—that we consult the man with the longest history in the company, Howard D. Woody. When asked what his father, John D. Woody, might have said about the past hundred years of the Union corporate adventure, Howard

said, "I think Dad would really be pleased with the progress." There has indeed been progress, and the great work of filling in all of the gaps in the service and linking neighbors together goes on.

John G. Woody suggested that Union's path towards maintaining longevity was to remain a local company, but offer service that worked everywhere. If the company provided good customer service and kept abreast of changing technology, he foresaw a long and prosperous future for Union Telephone. The family members and employees clearly recognize that reaching the centennial year is no reason to rest on one's laurels. Hard-working people who took on all of the challenges that both America's Outback and history had to throw at them made this century-long journey possible. There has been success and failure, growth and shrinkage, and highs and lows, but it always comes back to the teamwork and dedication of the Woody family and the Union extended family of loyal employees. As most rodeo riders will tell you, the ride is not over until you hit the ground.

Union's future is a misty ocean waiting for the forthcoming generations to pick up the oars, set a course, and begin the search for their own distant shores. John D. Woody's vision has piloted the company through its first century. As generation five and generation six take the helm, will they keep to the firm bearing set by their forbearers?

Wise people remind us that the future is not written in stone. The words painted on the huge boulder in front of Union's headquarters in Mountain View express a recollection of what has worked to get the company to its centennial year. Perhaps the sentiment on the boulder provides advice that will help future generations in their uncharted journeys yet to come. It is up to the coming generations to make that connection for themselves.

Union Telephone Board of Directors

J.C. (Cliff) Anderson
1950–1956

Dated records no longer exist that clearly identify early board members. To the best recollection of Howard D Woody (as told him by his father John D. Woody), these are the members of the earliest board of directors. These members served at some point between January 28, 1914 (when the company was formally established), and 1950, when Union began to keep dated records. It is interesting to note that at no time was John D. Woody president of the board of directors.

Ernest Dahlquist
1950–2004

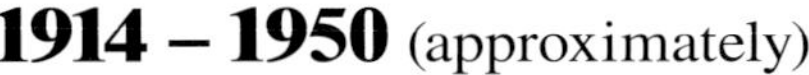

1914 – 1950 (approximately)

John Briggs
George Stoll Jr.
Tom Welch
George Smith
John D. Woody

Howard D. Woody
1950–

Board meeting minutes from 1950 to the present show the following board members and their periods of service to the company.

1950 – 1956

J. C. (Cliff) Anderson, President
Ernest A. Dahlquist, Secretary
Howard D. Woody, Board Member
Harry D. Buckley, Board Member
Albert H. Neff, Board Member

Harry D. Buckley
1950–1973

1956 – 1973

Howard D. Woody, President
Ernest A. Dahlquist, Secretary
Harry D. Buckley, Board Member
Albert H. Neff, Board Member
C. J. Dykes, Board Member

C.J.(Coldwell) Dykes
1956–2003

Albert H. Neff
1950–2004

1973 – 1991

Howard D. Woody, President
Ernest A. Dahlquist, Secretary
Albert H. Neff, Board Member
C. J. Dykes, Board Member
John G. Woody, Board Member

1991 – 2001

Howard D. Woody President
Ernest A. Dahlquist, Secretary
Albert H. Neff, Board Member
C. J. Dykes, Board Member
John G. Woody, Board Member
James H. Woody, Board Member
Gloria D. Woody, Board Member

2001 – 2003

Howard D. Woody, President
Gloria D. Woody, Secretary
Albert H. Neff, Board Member
C. J. Dykes, Board Member
John G. Woody, Board Member
James H. Woody, Board Member
Lee J. Woody, Board Member

2003 – 2004*

Howard D. Woody, President
Gloria D. Woody, Secretary
Albert H. Neff, Board Member
John G. Woody, Board Member
James H. Woody, Board Member
Lee J. Woody, Board Member
Ernest A. Dahlquist, Board Member

John G. Woody
1973–

James H. Woody
1991–

Gloria D. Woody
1991–2004

Lee J. Woody
2001–2007

Leslie F. Henderson
2004–

2004 – 2005

Howard D. Woody, President
Leslie F. Henderson, Secretary
Albert H. Neff, Board Member
John G. Woody, Board Member
James H. Woody, Board Member
Lee J. Woody, Board Member

Robert Brian Woody
2008–

2005 – 2008**

Howard D. Woody, President
Leslie F. Henderson, Secretary
Albert H. Neff, Board Member
John G. Woody, Board Member
James H. Woody, Board Member
F. Walter Riebenack, Board Member

Eric Woody
2008–

2008 – 2009

Howard D. Woody, President
John G. Woody, Vice President
James H. Woody, Treasurer
Leslie F. Henderson, Secretary
Albert H. Neff, Board Member
Robert Brian Woody, Board Member
Eric Woody, Board Member
Stacey (Woody) Aughe, Board Member
F. Walter Riebenack, Board Member

Stacey (Woody) Aughe
2008–

F. Walter Riebenack
2008–

2009 –

Howard D. Woody, President
John G. Woody, Vice President
James H. Woody, Treasurer
Leslie F. Henderson, Secretary
Ruth Barclay, Assistant Corporate Secretary
Robert Brian Woody, Board Member
Eric Woody, Board Member
Stacey (Woody) Aughe, Board Member
F. Walter Riebenack, Board Member

Ruth Barclay
2009–

Note: The board was re-staffed several times from 2003 to 2008 due to the untimely deaths of *Gloria D. Woody, 9/6/2004, and ** Lee J. Woody, 12/14/2007.

Union Telephone Highlights: A Hundred-and-One Years in a Glimpse

The following presents a brief history of Union Telephone's first hundred years. The entries in italics are not related to Union, but they provide markers as to what was happening in the history of the industry or the nation.

1914	John D. Woody founds the company from the Mountain View Mutual Telephone line, the Smith's Fork Mutual Telephone line, the Black Forks Telephone line, and the Lonetree-Linwood Telephone Company.
	First World War begins.
1915	Union Telephone moves into its first office building.
	First coast-to-coast long-distance service created by AT&T.
1916	John D. Woody goes to Sweet Home, Oregon, and establishes a new telephone company. Union Telephone operations are run by Earl Stoll.
1917	Earl Stoll builds the Lonetree to Mountain View line.
	The US telephone system is nationalized as a war measure.
	Isabell Woody passes away.
1918	John D. Woody returns to running Union Telephone's operations.
	First World War ends.
	Swine flu pandemic kills millions worldwide.
1919	Drought of 1919.
1920	*United States begins prohibition of alcoholic beverages with the passage of the Eighteenth Amendment of the US Constitution.*
1921	*First transcontinental flight by the US Post Office's Air Mail service.*
1922	Howard D. Woody is born.
	Teapot Dome scandal in Wyoming damages prestige of oil and gas industry.
1923	*First radiotelegraph message is sent.*
1924	Esther P. Gladwill (Woody) is born.
1925	*Scopes "monkey trial" (concerning the teaching of evolution) occurs.*
1926	*First public demonstration of television.*
1927	*The snowmobile is patented.*
1928	*A radiotelephone connection between the US and Europe is established.*
1929	Union Telephone connects to the AT&T "B" line and begins providing long-distance service.
	The Great Depression begins.
1930	*Construction of the Hoover Dam begins.*
1931	*Severe droughts create the "Dust Bowl."*

1932 *Lindbergh baby kidnapped.*

1933 *Twenty-Second Amendment to the Constitution of the United States repeals prohibition.*

1934 *Shirley Temple makes her film debut in Stand up and Cheer.*

1935 Goldie Woody passes away.

Franklin Roosevelt signs Executive Order 7037, creating a push for extending electrical service to rural areas.

1936 *The Rural Electrification Act creates the REA.*

1937 Azariah Woody passes away.

Trona is discovered in southwest Wyoming.

1938 *First public demonstration of color television.*

1939 *The Second World War begins.*

End of the Great Depression.

1940 Howard D. Woody enlists in the Wyoming National Guard.

1941 *The United States enters the Second World War.*

1942 *Radar technology is made operational.*

1943 Howard D. Woody marries Esther P. Gladwill.

1944 John G. Woody is born.

1945 *The Second World War ends.*

1946 Howard D. Woody returns from the Army.

Fort Bridger Telephone Company merges with Union in an exchange of stock.

1947 Gloria Dee Woody is born.

1948 The proposed sale of Union Telephone to AT&T/Mountain Bell is refused.

John D. and Mary E. Woody lease Union Telephone from the board of directors.

1949 *The Blizzard of 1949 hits Wyoming.*

The Union Telephone "snow plane" is used to haul supplies and mail to the community during the blizzard emergency.

James H. Woody is born.

1950 *North Korea invades South Korea, setting off the Korean conflict.*

1951 *The UNIVAC computer is introduced.*

1952 Lee Joe Woody is born.

1953 The Union board of directors stymies Howard D. Woody's attempt to apply for loan through the REA.

Union acquires the Bridger Telephone exchange in a stock deal.

1954 *Photovoltaic (solar) cells are invented.*

1955 *Commercial electricity from a nuclear power plant is put on the market for the first time.*

1956 Union acquires the Lyman Telephone Company in an exchange of stock.

Howard D. Woody becomes a member of the board of directors, the first time a Woody serves on the board.

Union receives a $300,000 loan through the Rural Electrification Administration.

1957 Union implements the dial system.

Union hooks up to the AT&T "B" cable near Urie.

Howard Woody becomes Union's general manager.

The switchboards for Mountain View and Manila are obtained from Leich Electric.

Union completes the Manila to Dutch John line.

Service to the US Bureau of Reclamation commences as the Flaming Gorge dam project begins.

Office buildings are constructed in Manila and Mountain View.

The board of directors approves applications for additional REA loans.

1958 Union contracts to remodel its old office building and build a new Dutch John exchange building.

Bonnie Jean Woody (Shannon) is born.

1959 Union seeks permission from the Public Services Commission to serve Granger and Seedskadee.

1960 The Dutch John to Brown's Park line is built.

Union's first year of profitability.

1961 *Humans orbit the earth for the first time.*

1962 Union's founder, John D. Woody, passes away.

Union acquires service territory in northwest Colorado.

Union installs its first improved mobile-telephone service.

1963 Union purchases an airplane for personnel transportation to distant and remote sites.

The first touch-tone telephone by Bell Systems is implemented in Pennsylvania.

1964 Howard Gene Woody is born.

Union provides radiotelephone service for the First Lady, Ladybird Johnson, during the dedication of the Flaming Gorge Dam.

1965 *The optical disc and hypertext are invented.*

1966 *Heavy smog kills over four hundred people in New York City.*

1967 *The SEACOM telephone cable begins service.*

1968 Stacey Ann Aughe is born.

The Brown's Park line is installed by contractors.

The first 9-1-1 service in the United States is implemented.

1969 James D. Woody and Eric Jon Woody are born.

Union commences service in Brown's Park and Christmas Meadows.

Union establishes a pension plan for its employees.

Neil Armstrong is the first human to walk on the moon.

1970 James H. Woody assumes corporate accounting duties.

1971 John G. Woody promoted to system engineer.

1972 Union upgrades to a one-party service and commences the plowing of cables for Bridger Valley.

Union upgrades the microwave system (analog) for its Utah exchanges.

Robert Brian Woody is Born.

Union receives an REA loan for systems upgrades.

1973 Union receives a loan from the Rural Telephone Bank for $1,375,500.
The first cellular telephone call is made in Manhattan.

1974 Union contracts for the construction of the Lyman Central Office building.

1975 Union receives a loan from the Rural Telephone Bank for $2,047,500.

1976 Union is among the first small telephone companies to install computerized billing.

1977 *The directory-assistance number, 4-1-1, is implemented by Bell systems.*

1978 Union purchases its current headquarters site.

1979 John G. Woody is injured in a single-car crash.
The first digital switch is installed in Bridger Valley.

1980 Roger Lind is promoted to Outside Plant Superintendent.

1981 *The IBM PC and the RCA Selectavision videodisc system are introduced.*

1982 John G. Woody is appointed Assistant Manager of Union.
Internet protocol suite (TC/IP) is introduced.

1983 Union headquarters moves to its current site.
Union installs upgraded switches in Dutch John and Greendale.

1984 AT&T is broken-up into "Baby Bells."
A digital main switch is installed at Christmas Meadows.
Union installs an operator switchboard in Mountain View.

1985 Operator services return to Union.
The company begins upgrading from analog to digital microwave.

1986 The microwave upgrade to digital is completed, except for US West paths.
The line between Hickey Mountain and Lonetree is plowed in.
Union purchases Angwin Cable Television, servicing Mountain View.

1987 Union agrees to buy and install a digital main switch with equal access, toll tandem facilities, and an integrated operator switching system.

1988 Union replaces the Dutch John to Brown's Park Line.
Union acquires bandwidth in Wyoming and Utah for wireless service.
Union replaces its analog system with a digital one.

1989 Union acquires bandwidth in Colorado.
The board of directors approves the construction of the cellular system.

1990 Union is loaned funding from CoBank to build cellular service.
Cellular service begins as Union Cellular.
Union opens its first storefront.
Lottiebelle Woody passes away.

1991 John G. Woody is elected to Union's board of directors and appointed vice president.

1992 Howard D. Woody resigns as general manager, and James H. Woody is appointed general manager.

1993 Union pursues acquisitions of the Afton, Elk Mountain, Encampment, Hanna, LaBarge, Rock River, Saratoga, and Shirley Basin exchanges.
Mary Ellen Woody passes away.

1994	Union borrows $30.5 million and uses it to acquire the exchanges at Elk Mountain, Encampment, Hanna, LaBarge, Rock River, Saratoga, and Shirley Basin.
1995	Stock dividends are issued.
	Union expands cellular service into Idaho.
1996	Union's bylaws are modified to establish a management team and to eliminate the positions of executive vice president and manager.
1997	Union issues an employee handbook, and 401K plans are offered to employees.
	Membership for the board of directors is set between five and nine members.
	Linda Kay Cox Woody passes away.
1998	The employee-compensation plan is approved by Union's board of directors.
1999	The company tops 31,000 cellular and 8,400 landline customers.
2000	*Y2K (Year 2000) scare.*
2001	James H. Woody and Gloria Dee Woody are added to the Union board of directors.
2002	Union expands its frequencies with the addition of 30MHz near Casper and 700MZz statewide.
2003	Union acquires Pyxis Communications and expands service statewide in Wyoming.
	Union acquires frequencies in Lincoln and Sweetwater counties from Edge Wireless.
2004	Esther P. Woody and Gloria Dee Woody pass away.
2005	Union Wireless replaces Union Cellular.
	James D. Woody passes away.
2006	Lottiebelle Woody passes away.
2007	Lee Woody passes away.
	Eric Woody, Robert Brian Woody, and Stacey Aughe are added to the management team.
2008	Eric Woody, Robert Brian Woody, and Stacey Aughe are added to board of directors.
	Union secures a $100 million loan for the expansion of its cellular system and the construction of Universal Mobile Telecommunications System.
2009	John G. Woody is appointed chief executive officer.
	James H. Woody becomes chief financial officer.
	Union offers roaming services in Europe.
	Union sells Wasatch Front 700Mhz frequencies to AT&T.
	Union's management team is modified in the company's bylaws to include a CEO, a CFO, and other positions.
2010	The Delaware Holding Corporation is established.
	Verla R. Woody passes away.
2011	Union receives the "Toughest Site Competition" award for its cellular towers.
	Union discontinues its cable-television service.
2012	Union receives letters of credit from CoBank to enter the Federal Communications Commission Mobility 1 auction.
2013	*Software Defined Networking attains practicality.*
2014	Union enjoys its centennial celebration.

REFERENCES

Bagley, Will. *Blood of the Prophets: Brigham Young and the Massacre at Mountain Meadows.* Norman, OK: University of Oklahoma Press, 2002.

Barker, W. H. "Operating Rural Lines." *Telephony* 19 (March 1910): 286

Bellis, Mary. "The History of the Electric Telegraph and Telegraphy." http://inventors.about.com/od/tstartinventions/a/telegraph.htm

Bigler, David L., and Will Bagley. *The Mormon Rebellion: America's First Civil War 1857–1858.* Norman, OK: University of Oklahoma Press, 2011.

Brooks, John. *Telephone: The First Hundred Years.* New York: Harper & Row, 1976.

Casson, Herbert Newton. *The History of the Telephone.* Project Gutenberg, 2008. http://www.gutenberg.org/files/819/819-h/819h-htm.

Coe, Lewis. *The Telephone and Its Several Inventors: A History.* Jefferson, NC: McFarland & Company, 1995.

Davis, Louise, and Nikki Walker. *Footsteps from the Past: Memories of Mountain View.* Mountain View, WY: Watkins Printing, 2004.

Dunham, Dick, and Vivian Dunham. *Flaming Gorge Country: The Story of Daggett County, Utah.* Denver, CO: Eastwood Printing and Publishing Company, 1993.

Elliot, Roy A. *Profiles of Progress: The Story of Early Events and Pioneer Families of Linn County, Oregon.* Self-published, 1991.

Fischer, Claude S. *America Calling: A Social History of the Telephone to 1940.* Berkeley, CA: University of California Press, 1992.

Garceau, Dee. *The Important Things of Life: Women, Work, and Family in Sweetwater County, Wyoming, 1880–1929.* Lincoln, NE: University of Nebraska Press, 1997.

Gardner, Dudley, and Verla R. Flores. *Forgotten Frontier: A History of Wyoming Coal Mining.* San Francisco: Westview Press, 1989.

Hamblin, Kathaleen Kennington. *Bridger Valley: A Guide to the Past.* Mountain View, WY: Self-published, 1993.

Hoegh, Leo Arthur, and Howard J. Doyle. *Timberwolf Tracks: The History of the 104th Infantry Division, 1942–1945.* Washington, DC: Infantry Journal Press, 1946.

Hope, John. "The Snyder Brothers" (West Texas Historical Association annual meeting, Lubbock, TX, April 1, 2011).

Larson, Thomas A. *History of Wyoming.* 2nd ed. Lincoln, NE: University of Nebraska Press, 1990.

Lee, John Doyle. *Mormonism Unveiled; or, The Life and Confessions of the Late Mormon Bishop, John D. Lee.* St. Louis, MO: Bryan, Brand, & Co., 1877.

MacDougall, Robert. *The Peoples Network: The Political Economy of the Telephone in the Gilded Age.* Philadelphia: University of Pennsylvania Press, 2014.

Moulton, Candy. *Roadside History of Wyoming.* Missoula, MT: Mountain Press Publishing Company, 1995.

Owen, James P., and David R. Stoecklein. *Cowboy Ethics: What Wall Street Can Learn from the Code of the West.* Guilford, CT: Lyons Press, 2005.

Standage, Tom. *The Victorian Internet.* New York: Walker House, 1998.

Stewart, Elinore Pruitt. *Letters of a Woman Homesteader.* New York: Dover Publications, Inc., 1914.

Tittsworth, William G. *Outskirt Episodes.* Des Moines, IA: Success Composition and Printing Company, 1927.